NANOTECHNOLOGY SCIENCE AND TECHNOLOGY

MOLECULAR DYNAMICS OF NANOBIOSTRUCTURES

NANOTECHNOLOGY SCIENCE AND TECHNOLOGY

Additional books in this series can be found on Nova's website
under the Series tab.

Additional E-books in this series can be found on Nova's website
under the E-book tab.

NANOTECHNOLOGY SCIENCE AND TECHNOLOGY

MOLECULAR DYNAMICS OF NANOBIOSTRUCTURES

KHOLMIRZO KHOLMURODOV
EDITOR

Nova Science Publishers, Inc.
New York

NOTICE TO THE READER

The Publisher has taken reasonable care in the preparation of this book, but makes no expressed or implied warranty of any kind and assumes no responsibility for any errors or omissions. No liability is assumed for incidental or consequential damages in connection with or arising out of information contained in this book. The Publisher shall not be liable for any special, consequential, or exemplary damages resulting, in whole or in part, from the readers' use of, or reliance upon, this material. Any parts of this book based on government reports are so indicated and copyright is claimed for those parts to the extent applicable to compilations of such works.

Independent verification should be sought for any data, advice or recommendations contained in this book. In addition, no responsibility is assumed by the publisher for any injury and/or damage to persons or property arising from any methods, products, instructions, ideas or otherwise contained in this publication.

This publication is designed to provide accurate and authoritative information with regard to the subject matter covered herein. It is sold with the clear understanding that the Publisher is not engaged in rendering legal or any other professional services. If legal or any other expert assistance is required, the services of a competent person should be sought. FROM A DECLARATION OF PARTICIPANTS JOINTLY ADOPTED BY A COMMITTEE OF THE AMERICAN BAR ASSOCIATION AND A COMMITTEE OF PUBLISHERS.

Additional color graphics may be available in the e-book version of this book.

LIBRARY OF CONGRESS CATALOGING-IN-PUBLICATION DATA

Molecular dynamics of nanobiostructures / editor, Kholmirzo Kholmurodov.
   p. ; cm.
 Includes bibliographical references and index.
 ISBN 978-1-61324-320-6 (hardcover)
 1. Nanostructures. 2. Biotechnology. 3. Molecules--Models. 4. Biomimetic materials. I. Kholmurodov, Kholmirzo.
 [DNLM: 1. Nanostructures. 2. Models, Molecular. QT 36.5]
 TP248.25.N35M65 2011
 660.6--dc23
                          2011012974

Published by Nova Science Publishers, Inc. † New York

MOLECULAR DYNAMICS OF NANOBIOSTRUCTURES

Proceedings of the 4th Japan-Russia International Workshop MSSMBS' 10
"Molecular Simulation Studies in Material and Biological Sciences"

JINR-MSU, Dubna-Moscow, Russia, September 26-29, 2010

KHOLMIRZO KHOLMURODOV
(EDITOR)

Center for Molecular Modeling, Laboratory of Radiation Biology
Joint Institute for Nuclear Research, 141980 Dubna, Moscow Region, Russia

E-mail address: mirzo@jinr.ru

DATE AND PLACE: September 26-29, 2010, Dubna, Moscow Region, Russia

ORGANIZING COMMITTEE: Kholmirzo Kholmurodov (Co-Chairman), Alexander Nemukhin (Co-Chairman, MSU), Valentina Novikova (Scientific Secretary, JINR International Cooperation Department), Viktor Krylov, Sergei Negovelov, Mikhail Altaisky, Guzel Aru, Gennady Timoshenko, Olga Mashkova

ORGANIZED BY: The Laboratory of Radiation Biology of the Joint Institute for Nuclear Research and the Chemistry Faculty of Moscow State University

TOPICS COVERED:

- Novel MD simulation techniques & methods

- Hybrid computational approaches: DFT, QM/MM, MD, MD/CFD

- Novel computing & and communication architectures

- General- & special-purpose MD machines

- Video-game computers for accelerating MD

- Simulation of biomacromolecules

- Protein & DNA modelling

- Simulation of radiation-induced mutations

- Simulation of crystal & polymer systems

- Quantum biophysics, electronic structure of macromolecules

THE PRINCIPAL AIMS: We have provided a platform for computer molecular simulation communities and scientists working in the material and biological areas to meet and share thoughts on latest trends. The main objective was to focus on the molecular dynamics simulations of chemical physics and biophysical systems.

LIST OF CO-SPONSORS:

- JINR (Joint Institute for Nuclear Research)

- Moscow Medical Center "Avicenna", Ltd.

- RFBR (Russian Foundation for Basic Research)

- LIT - Grid NNS (Laboratory of Information Technologies, JINR)

Joint photo of the International Workshop MSSMBS' 10

From left to right:
Gleb Sitnikov, Mikhail Altaisky, Nina Shmakova, Leonid Chernozatonskii, Tamara Fadeeva, Viktor Krylov, Larisa Melnikova, Albina Kireeva, Mikhail Kiselev, Tatyana Feldman, Roman Efremov, Aram Shahinyan, Fuyuki Shimojo, Alexander Didyk, Kholmirzo Kholmurodov, Satoshi Ohmura, Vladimir Korenkov, Oleg Titov, Dmitrii Lakhno, Alexander Korotaev, Kenji Yasuoka, Iori Yonekawa, Alexander Vorontsov, Vladimir Stegailov, Tomoyuki Yamamoto, Yuko Okamoto, Genri Norman, Dmitrii Shulga, Daisuke Murakami

CONTENTS

PREFACE

A lot of discoveries in modern science and technology (in particular, recent progress in nanotechnologies) are inseparably linked with the use of computer molecular simulation methods. Today, molecular simulation is one of the basic instruments in exploring the properties of nano- and biostructures. Computer molecular simulation is a powerful technique to investigate various physical or biological processes at the atomic level. Molecular simulation is a practical tool for the development of new materials and new drugs, as well as for performing large-scale calculations on molecular complexes of hundreds of thousands or multi-million particle systems. With the creation of new parallel/vector supercomputers and special-purpose computational clusters, molecular simulation methods are becoming a real tool in bioengineering, nanotechnology, and materials science, capable of estimating the details of the atomic-scale processes and technologically interesting phenomena.

Computer molecular simulation is a set of molecular simulation and quantum chemistry methods, or hybrids of these two kinds of methods exhibiting new possibilities. The methods of computer molecular simulation (conventional or hybrid molecular dynamics (MD), Monte-Carlo, *ab initio* quantum-chemistry, etc.) of large molecular systems, which were first proposed more than 50 years ago with the invention of computers, have been rapidly developing in the last 5–10 years. Molecular simulation (conventional and hybrid MD) is based on classical Newtonian physics, modeling the particle interaction in molecules via the force fields defined in advance – empirically or calculated by other methods. In a MD study, the molecular systems are modeled deterministically by the integration of classical equations of motions; in MC, stochastically – with various ensembles. The MD methods are capable of modeling atomic molecular systems of up to thousands and million particles and simulating many system parameters and environmental configurations. MD simulation allows one to predict efficiently the ensemble properties and behavior, such as

P–V–T relations, phase equilibrium, transport properties, structures of synthetic and biological macromolecules, docking of one molecule against another, etc.

Computational quantum chemistry research – *ab initio*, density functional theories (DFT), and others – in contrast to the conventional molecular simulations, is based on quantum physics. The computational quantum chemistry methods were first applied to the electronic structure of atoms or molecules, which yielded wave functions or a probability density functional describing the electron states. The quantum chemistry methods provide greater accuracy but are restricted to a smaller molecular size because of their complexity and CPU costs. Quantum chemistry simulation is essential when chemical bonds are formed or broken. It is also used when force parameters are unknown or not applicable. The DFT methods are well established and used with increasing accuracy; the high-level wave function methods with large atomic orbital basis sets currently remain standard. As results of our quantum chemistry studies, we got force – field data, which enabled us to calculate the thermochemical, kinetic, and optical properties, NMR shifts, etc.

In this book, original papers are collected that demonstrate efficient uses of molecular dynamics (MD) simulation for studying nanoscale phenomena in a number of models from material and life sciences. This volume contains the Proceedings of the 4th Japan–Russia International Workshop "Molecular Simulation Studies in Material and Biological Sciences" (MSSMBS'2010), which was held at the Joint Institute for Nuclear Research (JINR), Dubna, and at the Chemistry Faculty of Moscow State University on September 26–29, 2010. The series of MSSMBS meetings started in 2004 and was continued in 2006, 2008, and 2010. MSSMBS'2004 was the first international conference held in Russia that was focused on methodological problems and applications of the art of MD simulations in physical, chemical, and biological systems. The MSSMBS workshops are mostly contributed by leading research groups of Japan and Russia and are also participated by scientists of European institutes. The subjects of the MSSMBS workshops include different aspects of molecular simulation in materials science and biological research; computational and theoretical studies of atomic and molecular interactions; dynamics between atoms, molecules, ions, clusters, and surfaces; modern high-performance computing facilities; and simulation techniques and methods applied for studying molecular systems and structures.

The scope of the MSSMBS'10 meeting included, in particular, the following topics:

- Novel MD simulation techniques & methods
- Hybrid computational approaches: DFT, QM/MM, MD, MD/CFD
- Novel computing & and communication architectures
- General- & special-purpose MD machines
- Video-game computers for accelerating MD
- Simulation of biomacromolecules
- Protein & DNA modelling
- Simulation of radiation-induced mutations
- Simulation of crystal & polymer systems
- Quantum biophysics; electronic structure of macromolecules

    The MSSMBS'10 workshop started at the JINR International Conference Hall, Dubna, and then continued and finished at the Chemical Faculty of Moscow State University. As the MSSMBS International Workshops are focused on computer molecular simulation, which is a booming area, they provide an excellent opportunity for members of leading research teams of Japan, Russia, and Europe to meet and share their thoughts on the latest trends in this art. The MSSMBS workshops also aim to promote a new field of science at JINR, Dubna, which has traditionally been a nuclear research center. We held a broad discussion on results of radiobiological and nuclear physics investigations conducted at JINR's basic experimental facilities.

Kholmirzo T. Kholmurodov
Chairman of the MSSMBS
Organizing Committee

JINR (Joint Institute for Nuclear Research)
6 Joliot-Curie St., Dubna, 141980, Moscow Region, Russia
E-mail: mirzo@jinr.ru

MOLECULAR SIMULATIONS IN MATERIAL AND BIOLOGICAL RESEARCH: AN INTRODUCTION

On September 26–29, the Laboratory of Radiation Biology of the Joint Institute for Nuclear Research and the Chemistry Faculty of Moscow State University successfully conducted the 4th Japan–Russia International Workshop MSSMBS-2010 "Molecular Simulation Studies in Material and Biological Sciences" (co-chaired by Profs. Kholmirzo T. Kholmurodov and Alexander V. Nemukhin). The MSSMBS-2010 scientific program provided a lot of interesting talks on different aspects of materials science and biological research and outlined the current state and prospects of computer molecular simulation studies; its covered a broad spectrum of problems in modern physics, biochemistry, and nanotechnology. In particular, the following topics were discussed: first-principles calculations and synchrotron radiation analysis of functional materials; hydrated lipid bilayers; DNA nano-bioelectronics; graphene-graphane mixing nanostructures; generalized-ensemble simulations in protein science; the atomic theory of nucleation in solids and liquids; specific features of MD simulations of peptides, proteins, and membranes; MD simulations of human red blood cells; studies of nematic liquid crystals; density-functional-based molecular dynamics simulations of disordered materials; ab initio models of electronic states; and so on. The MSSMBS-2010 participants represented leading computer molecular simulation research groups of Japan (Waseda University, Nagoya University, Keio University, Kumamoto University, etc.) and Russia (the Institute of Mathematical Problems of Biology, Pushchino Scientific Center, M.V. Lomonosov Moscow State University, Emanuel Institute of Biochemical Physics, RAS, etc.).

In the first paper by T. Yamamoto and K. Kawabata local environments of dopants in functional ceramic materials synthesized by the solid-state reaction method are systematically investigated by using the X-ray absorption near-edge structure (XANES) analysis with the aid of the first-principles calculations. Four types of materials, (1) bioceramics, (2) dilute magnetic oxide, (3) phosphor and (4) electrolyte of solid fuel cell, are chosen to be examined here. This analytical method by combined use of XANES and first-principles calculations has successfully explained the local environment of dopants in the above materials in an atomic scale.

In the the second paper by T. Feldman, Kh. Kholmurodov and M. Ostrovsky the computer simulation results are discussed with the actual role that 11-*cis*-retinal plays as a chromophore group which enables rhodopsin for the ultra-fast and efficient photoisomerization and as a ligand-antagonist that keeps rhodopsin as a G-protein-coupled receptor in an inactivated state. Rhodopsin protein is a typical member of G-protein-coupled receptor (GPCR) family. It is the only receptor from a wide variety of the GPCRs for which a tertiary structure is known. In this paper the computer MD (molecular dynamics) simulation results of rhodopsin (with 11-*cis*-retinal chromophore) are presented. It is shown that 11-*cis*-retinal chromophore to be rearranged, after its insertion into the chromophore pocket. Namely, the beta-ionone ring of 11-*cis*-retinal is twisted in time frame of 0.4 ns starting from the simulation run. The changes in behavior of the nearest amino acid residues, surrounding the chromophore retinal, correlate with the beta-ionone ring twist. A clear correlation is observed between the beta-ionone ring twist and the mobility of the cytoplasmic domain that is responsible for binding of G-protein and stabilization of alpha-helix H-VI which is a characteristic site for the dark rhodopsin. The computer simulation results are discussed with the actual role that 11-*cis*-retinal plays as a chromophore group which enables rhodopsin for the ultra-fast and efficient photoisomerization and as a ligand-antagonist that keeps rhodopsin as a G-protein-coupled receptor in an inactivated state.

In the third paper by A. Shahinyan, P. Hakobyan and A. Poghosyan a 80 ns Molecular Dynamics (MD) simulation of human red blood cells (erythrocyte) asymmetric model membrane has been performed and the dynamical and structural properties have been investigated. The system consists of all major phospholipids found in human erythrocyte membrane by real experiment, cholesterol molecules and transmembrane part of Glycophorin A (GpA): an important intrinsic erythrocyte membrane protein. The surrounding molecular composition of the protein was investigated and the influence of surrounding phospholipid and cholesterol molecules on protein properties was discussed. The NAMD code with corresponding CHARMM27 force field was used. Some structural parameters of membrane, such as area per molecule, membrane thickness, surface roughness, etc were discussed.

In the 4th paper by Y. Mori, A. Mitsutake and Y. Okamoto generalized-ensemble simulations in protein science are developed. In simulations in materials and biological sciences, one encounters with a great difficulty that conventional simulations will tend to get trapped in states of energy local minima. A simulation in generalized ensemble performs a random walk in potential energy space and can overcome this difficulty. From only one simulation run, one can obtain canonical-ensemble averages of physical quantities as functions of temperature by the histogram reweighting techniques. The author review the generalized-ensemble algorithms. The multidimensional extensions of the replica-exchange method and simulated tempering are presented. The effectiveness of the methods is tested with short peptide and protein systems.

In the 5th paper by D. Pyrkova, N. Tarasova, N. Krylov, D. Nolde, and R. Efremov the studies of lateral heterogeneity in cell membranes have been performed which are important since they help to understand the physical origin of lipid domains and rafts. The simplest membrane mimics are hydrated bilayers composed of saturated and unsaturated lipids. While their atomic structural details resist easy experimental characterization, important insight can be gained *via* computer modeling. The author present the results of all-atom molecular dynamics simulations for a series of fluid one- and two-component hydrated lipid bilayers composed of phosphatidylcholines with saturated (dipalmitoylphosphatidylcholine, DPPC) and mono-unsaturated (dioleoylphosphatidylcholine, DOPC) acyl chains slightly differing in length (16 and 18 carbon atoms, respectively). As a results, it was shown that the bilayers' properties are tuned in a wide range by the chemical nature and relative content of lipids. The impact that the micro-heterogeneity may have on formation of lateral domains in response to external signals is discussed. Understanding of such effects creates a basis for rational design of artificial membranes with predefined properties.

The 6th paper by A. Sevenyuk, V. Blinov, V. Golo and K. Shaitan presents a study on the cholesteric liquid crystallin phases of the DNA, and a computer simulation that relies on the electrostatical picture of the interaction between molecules of the DNA in solution. To that end the authors use a qualitative model that describes a molecule of the DNA as a one-dimensional lattice framed by charges due the phosphate groups, and dipoles of base pairs subject to the helicoidal symmetry. The results are in agreement, by orders of magnitude, with the experimental data.

In the 7th paper by S. Ohmura and F. Shimojo the atomic dynamics of liquid B_2O_3 and the energy-transfer mechanism in light harvesting dendrimers are studied by *ab initio* molecular-dynamics simulations. In liquid B_2O_3, it is found that the diffusivity of boron becomes about two times larger than that of oxygen under pressure above 20 GPa while the former is 10-20 % smaller than the latter at lower pressures. The authors reveal the microscopic origin of this anomalous pressure dependence of diffusivity. The nonadiabatic effects in molecular-dynamics simulations to describe the photo-exitation state of dendrimers have been taken into account. THe simulation reveals the key role of thermal molecular motion that significantly accelerates the energy transport based on the Dexter mechanism.

The 8th paper by A. Artyukh, L. Chernozatonskii, V. Artyukhov and P. Sorokin reviews the experimental and theoretical works on composite materials based on carbon nanotubes and graphene. Simultaneous use of 1D and 2D nanoparticles facilitates the linking between components, improving the mechanical and conductive properties of the resulting composite films in comparison with pure components. Such materials are promising for diverse applications including transparent electrodes.

In the 9th paper B. Grigorenko, M. Khrenova, J. Zhang and A. Nemukhin report the results of molecular dynamics simulations and quantum chemical calculations of the structure of light harvesting complex LH1 from the bacterial photosynthetic center of *Thermochromatium tepidum*. Structure and function of the light harvesting complex LH1 of the bacterial photosynthetic center of a thermophilic purple sulfur bacterium, *Thermochromatium tepidum*, present a challenge in studies of photoreceptor biomolecular systems since the experimental spectroscopy information is not currently augmented by the available crystallography data. By using the primary sequence of amino acid residues of the α- and β-polypeptide helices from this LH1 complex and the related templates the author constructed the three-dimensional structural model of the entire antenna system. Molecular dynamics simulations of this system solvated in water both for calcium free and calcium bound units show that the calcium ions can be trapped at least at two different binding sites. Quantum chemical calculations of the bacteriochlorophyll absorption bands at the predicted LH1 geometry configurations indicate that one of these sites is preferable to explain the observed red shifted absorption upon calcium binding.

The 10th paper by D. Murakami and K. Yasuoka discusses an ice nucleation protein that induces a phase transition from liquid water to ice in the air. A specific hydrophilic surface of the protein may have an influence on the network of hydrogen bonds between water molecules adsorbing onto the protein. However, microscopic characteristics of the ice nucleation protein and the behavior of water molecules on the protein have not been clarified. Therefore, molecular dynamics simulations of a system consisting of water and an ice nucleation protein was used to clarify some dynamics in the atomic level. As a result, there were some differences between simulation predict ions of water clusters adsorbed on the ice nucleation protein and the conventional percolation theory. It was found that finite clusters tend to be localized on the surface and trapped by sites of the protein. The initial results suggested the need for study on another type of hydrophilic protein and weaker hydrophilicity. The results pointed out the fact that the hydrophilicity of the ice nucleation protein influenced the formation of the water network that water clusters adsorbed on the ice nucelation protein tend to be localized.

In the 11th paper by Kh. Kholmurodov, E. Krasavin, V. Krylov, E. Dushanov, V. Korenkov, K. Yasuoka, T. Narumi, Y. Ohno, M. Taiji and T. Ebisuzaki, based on MD (molecular dynamics) simulation method a comparative analysis has been performed for the p53 dimer + DNA interaction of wild-type and mutant Arg273His (R273H) proteins. The aim of this paper is to study the molecular mechanism of the onco-protein p53 and DNA binding. A comparative analysis show that R273H mutation causes an essential effect on the p53–DNA interaction, removing their close contact. The obtained MD simulation results illustrate a detailed molecular mechanism of conformations of key amino acids in the p53–DNA binding domain, which is important for the physiological functioning of the p53 protein and understanding the origin of cancer.

In: Molecular Dynamics of Nanobiostructures   ISBN: 978-1-61324-320-6
Editor: K. Kholmurodov   © 2012 Nova Science Publishers, Inc.

Chapter 1

COMBINED USE OF THE SYNCHROTRON RADIATION AND THE FIRST-PRINCIPLES CALCULATIONS FOR THE LOCAL ENVIRONMENT ANALYSIS OF DOPANTS

Tomoyuki Yamamoto[1] *and Kazuhiko Kawabata*[2,*]
[1]Faculty of Science and Engineering, Waseda University
Shinjuku, Tokyo 169-8555, Japan;
[2]Department of Holistic Human Sciences, Kwansei Gakuin
University, Nishinomiya, Hyogo 662-8501, Japan

Abstract

Local environments of dopants in functional ceramic materials synthesized by the solid-state reaction method are systematically investigated by using the X-ray absorption near-edge structure (XANES) analysis with the aid of the first-principles calculations. Four types of materials, (1) bioceramics, (2) dilute magnetic oxide, (3) phosphor and (4) electrolyte of

*E-mail addresses: tymmt@waseda.jp. Waseda University, 3-4-1 Okubo Shinjuku-ku, Tokyo 169-8555, Japan; Tel+Fax: +81-3-5286-3317 (Corresponding author)

solid fuel cell, are choaon to be cxamined here. This analytical method by combined use of XANES and first-principles calculations has successfully explained the local environment of dopants in the above materials in an atomic scale.

Keywords: X-ray absorption near edge structure, First-principles calculation, Dilute magnetic semiconductor, Phosphor, Solid fuel cell, Bioceramics

1.1. Introduction

Doping technique is widely used to give an additional properties, e.g., electronic, magnetic and optical properties, etc., especially in semiconducting materials, which has been also applied recently to other types of functional materials. It is essential to know the local environment of dopants in an atomic scale, in order to understand the mechanism of newly appeared properties due to the doping, and to design new materials with desired properties by doping technique. There are some analytical methods to investigate such local environment of dopants, in which X-ray absorption near edge structure (XANES) is one of the most powerful methods. It was reported that XANES technique enables us to determine the local environment of dopant at an ultra dilute concentration level [1]. Conventional analysis of XANES is based on the fingerprint type method, in which experimental XANES spectrum of interest is compared with the experimental spectra of reference materials. However, we often meet difficulty to determine the local environment of dopant by such experimental fingerprint type analysis. To overcome this difficulty, a lot of attempts have been carried out to reproduce the experimental XANES profiles by theoretical calculations. If quantitative agreement can be obtained between experimental and calculated XANES spectra, we can get the theoretical fingerprints. For these theoretical XANES calculations, some kinds of calculating methods, i.e., molecular orbital method, band structure method and multiple scattering method [2], were employed. However, the theoretical spectra do not always reproduce the experimental spectra satisfactorily. There are several factors contributed to the poor agreement. One of the most typical reasons is improper treatment of the interaction between a core-hole and an excited electron, i.e., the core-hole effect. Proper inclusion of the core-hole effect is mandatory for reproducing experimental spectra by theoretical calculations. Recently, quantitative reproductions of experimental XANES from many different kinds of materials have been reported [3,4] by the first-principles band-structure calculations within the density functional theory (DFT) using the orthogonalized linear combination of the atomic orbital (OLCAO) method [5] and the full-potential augmented plane wave plus local orbitals (APW+lo) method [6]. In these calculations, the core-hole effect was directly included in the self-consistent calculations. Interaction among core-holes was minimized using supercells. Thereby the core-hole effect was included within the framework of one-electron approximation.

In the present study, we have summarized the local environment analysis of dopants in four kinds of materials, i.e., 1) bioceramics (Mn-doped β-$Ca_3(PO_4)_2$), 2) dilute magnetic semiconductor (Mn- and Fe-codoped In_2O_3), 3) phosphor (Pr-doped $SrTiO_3$ and $CaTiO_3$, Pr- and Ga-codoped $SrTiO_3$) and 4) electrolyte of solid fuel cell (Y-doped CeO_2), by the XANES analysis with the aid of the first-principles calculations. For these target subjects written above by 1)-4),Mn-K, Mn-K and Fe-K, Pr-L_3 and Ga-K, and Y-L_3 XANES were investigated here, respectively.

1.2.   Experimental and Computational Procedures

All the samples were prepared by the conventional solid-state reaction method. Prior to the XANES measurements, all the samples were characterized by using the X-ray diffraction (XRD) technique to check whether the synthesized material is single phase or not. All the present sample powders were determined to be a single phase by the XRD measurements.

Two types of XANES measurements, i.e., hard and soft X-ray XANES measurements, were carried out. Former XANES spectra, i.e., hard X-ray XANES, were collected at the beamlines in two synchrotron facilities, i.e., BL01B1 in SPring-8 and BL15 in SAGA Light Source with the transmission mode. Synchrotron radiations from the storage rings were monochromatized by the Si(111) or Si(311) double-crystal monochromators. Later soft X-ray XANES measurements, only Y-L_3 XANES in the present paper, were carried out at BL-1A in UVSOR by the total electron yield method, in which InSb(111) double-crystal monochromater was employed. The sample powders were mounted on the carbon adhesive tape, which were settled on the first dinode of the electron multiplier.

The first-principles calculations were carried out to obtain the theoretical XANES spectra by using the full-potential Augmented Plane Wave plus local orbital (APW+lo) package (WIEN2k [6]). As it is well established, the introduction of the core-hole effect is mandatory to reproduce the spectral fine structure of XANES [4]. In the present study, core-hole effect is introduced by removing one core electron of interest and putting one additional electron at the bottom of the conduction band, which approximately corresponds to the final state of the X-ray absorption process of interest. Muffin-tin radius of each atom was chosen as large as possible in the cell, and $R_{MT}K_{max}$, which corresponds to the plane wave cutoff, was set by using the product of the smallest R_{MT} and K_{max} (3.0 - 3.5 $Ry^{1/2}$). The Mokhorst-Pack scheme [7] was employed for k-point sampling in the reciprocal space, in which VN is set to be larger than 10000, where V and N are volume of the cell in $\mathring{A}^3$ unit and number of k-point, respectively. Here the supercell was used to minimize the interaction between core-holes due to the three dimensional periodic boundary condition in our band-structure calculations. Transition energy is calculated by a difference of total electronic energies between initial (ground) and final (core-holed) states. Spectral profiles were calculated by the product of projected partial density of state for the electric dipole allowed transition of interest and the radial part of the transition probability.

1.3. Results and Discussions

1.3.1. Bioceramics

Bioceramic materials have been extensively studied because of their potential applications in the medical field, e.g., dentistry and orthopedics. For such applications, bioactive ceramic materials based on hydroxyapatite (HAp) and calcium phosphates are employed to repair bone defects. Among these bioactive materials, β-tricalcium phosphate (β-TCP) is a potential candidate for bone substitution. Many attempts have been made to tune the resorption rate of bioceramics implanted for bone reforming by doping them with different types and/or concentrations of trace elements. To understand the influence of dopants, it is essential to know the local environment of the dopants in an atomic scale. In the present study, the substitution mechanism of Mn ions in β-TCP was directly investigated by performing a combination of XANES measurements and first-principles calculations[8].

Commercially available high purity powders of $CaHPO_4$, $CaCO_3$ and $MnCO_3$ were used as starting materials. To remove the hydrated water, $CaHPO_4$ and $CaCO_3$ powders were dried in the air at 473 K and 773 K, respectively, for

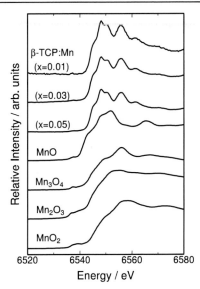

Figure 1.1.   Observed Mn K-edge XANES spectra of Mn-doped β-TCP, $Ca_{1-x}Mn_x(PO_4)_2$. From top to bottom, x = 0.01, 0.03, 0.05; and MnO, Mn_3O_4, Mn_2O_3, and MnO_2.

30 min. prior to weighing. Resulting powders were weighed changing the Mn concentrations of x = 0, 0.01, 0.03, 0.05 in $Ca_{3-x}Mn_x(PO_4)_2$, which were mixed and ground in an agate mortar for 30 min. and calcined at 1273 K for 6 hours in air. The observed Mn K-edge XANES spectra of Mn-doped β-TCP are shown in Fig. 1 together with those of Mn oxides with different oxidation states of Mn ions. The spectral fine structures of Mn-doped β-TCP with different concentrations of Mn ions showed almost the same features, but they were clearly distinguishable from reference Mn oxides. These results indicate that the local environment of Mn ions in β-TCP are nearly identical, when the Mn concentrations, x, vary from 0.01 to 0.05, but that they differ from those in Mn oxides. It is very difficult to determine the local environment of doped ions using the conventional fingerprint type analysis for this type of doped ion. Accordingly, the first-principles calculations were mandatory to obtain the theoretical Mn K-edge XANES, i.e., theoretical fingerprints, of the Mn ions substituted in β-TCP.

Prior to comparison of the observed and calculated XANES spectra of Mn-doped β-TCP, the validity of the present XANES calculation was examined by

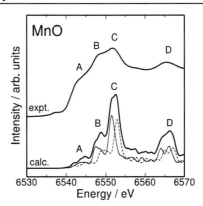

Figure 1.2. Comparison between observed and calculated XANES spectra of MnO at Mn K-edge. The transition energy of the calculated NEXAFS spectrum is corrected by $\Delta E = -20.7$ eV ($\Delta E/E = 0.3\%$).

comparing the calculated Mn K-edge spectrum of MnO to the experimental one, as shown in Fig. 2. This XANES calculation was conducted using the 2 x 2 x 2 supercell of a rock-salt structured conventional unit cell of MnO consisting of 64 atoms. The observed spectral fine structure of the Mn K-edge XANES of MnO was quantitatively well reproduced by the present calculations when the transition energy was corrected by $\Delta E = -20.7$ eV ($\Delta E/E = 0.3\%$). The calculated XANES spectra of Mn ions substituted for Ca(5) site, which is the energetically most favorable substitution model estimated by the plane-wave basis first-principles calculations, is compared with the experimental one in Fig. 3, in which the transition energy was corrected by the same amount of energy as for MnO, i.e., $\Delta E = -20.7$ eV. As shown in this figure, the experimental XANES profile has been quantitatively well reproduced by the calculated XANES spectrum, which shows doped Mn ions are substituted at Ca(5) site.

1.3.2.   Dilute Magnetic Oxide

Semiconductor doped with dilute magnetic elements, which are called as Dilute Magnetic Semiconductor (DMS), have been extensively studied since the discovery of carrier-induced ferromagnetism in $In_{1-x}Mn_xAs$ and $Ga_{1-x}Mn_xAs$ [9, 10]. After these discoveries, many attempts to increase the Curie temperature were made, which yielded room temperature ferromagnetism in DMS. Most of

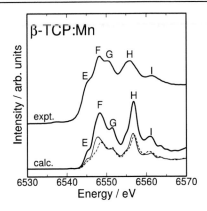

Figure 1.3. Comparison between observed and calculated XANES spectra of Mn-doped β-TCP at Mn K-edge. The transition energy of the calculated NEXAFS spectrum is corrected by $\Delta E = -20.7$ eV ($\Delta E/E = 0.3\%$).

these researches of DMS have been limited using the thin-films, but recently it was reported that the dilute magnetic oxide (DMO) In_2O_3:(Mn, Fe) showed the room temperature ferromagnetism [11]. Then the local environment of the magnetic elements, i.e., Mn and Fe ions, in In_2O_3 were examined here. Sample specimen, $(In_{0.94}Mn_{0.03}Fe_{0.03})_2O_3$, was fabricated by using the commercially available high purity powders of In_2O_3, $MnCO_3$ and Fe_2O_3. After these powders were mixed and ground in an agate mortar, they were pressed into a pellet form and sintered in air for 12 hours at 1423 K. Prior to the XANES measurements, magnetic property of this specimen was examined by the superconducting quantum interference device (SQUID) and the vibrating sample magnetometer (VSM), which are shown in Fig. 4(a) and (b). From the results of these experiments, the room temperature ferromagnetism in this specimen was confirmed and the Curie temperature was estimated at around 800 K. Observed Mn-K and Fe-K XANES spectra of In_2O_3:(Mn, Fe) are shown in Fig. 5 (a) and (b) together with the reference Mn and Fe oxides, respectively. Both of the spectral profiles of In_2O_3:(Mn, Fe) show different features from those of the reference oxide materials, which indicate the local environment of Mn and Fe ions are different from those of these reference oxides. From the results of XRD and these comparisons of XANES spectra, it is suggested that both of the Mn and Fe ions are substituted in In_2O_3. In order to confirm this suggestion, the

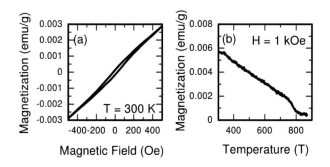

Figure 1.4. (a) Magnetization vs magnetic field curve of In_2O_3:(Mn, Fe) at 300 K by using SQUID and (b) temperature dependence of magnetization at 1 kOe by using VSM.

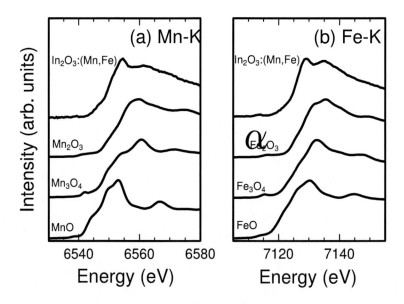

Figure 1.5. Observed (a) Mn-K and (b) Fe-K XANES spectra of In_2O_3:(Mn, Fe) and reference oxide maerials.

first-principles calculations were performed to obtain the theoretical XANES spectra. At first, XANES spectra of the reference Mn and Fe oxides were calculated to check the accuracy of the calculations. Here MnO and Fe_3O_4 were chosen to be calculated. Cell sizes of the present calculations are 2x2x2 supercell of conventional rock-salt structured MnO (64 atoms) as emloyed in the previous section and unit cell of spinel structured Fe_3O_4 (56 atoms). Spin-polarized calculations were carried out for both cases. Resultant theoretical XANES spectra of MnO and Fe_3O_4 are compared with the experimental ones in Fig. 6 (a) and (b), respectively. Characteristic features of the experimental Mn-K and Fe-

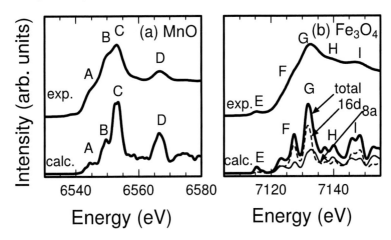

Figure 1.6. Comparison of XANES spectra of (a) Mn-K of MnO and (b) Fe-K of Fe_3O_4 between experiments and calculations. Thin solid and dashed lines denote calculated spectra from Fe at 16d and 8a sites, respectively, and thick solid line is the sum of these two in (b). Calculated transition energies are corrected by (a)ΔE_{Mn} = -19.5 eV and (b)ΔE_{Fe} = -20.0 eV.

K XANES spectra are well reproduced by the present calculations, when the transition energy were corrected by ΔE_{Mn} = -19.5 eV ($\Delta E/E$ = 0.3%) and ΔE_{Fe} = -20.0 eV ($\Delta E_{Fe}/E$ = 0.3%) for Mn and Fe K-edges, respectively. Then the Mn and Fe substituted models in In_2O_3 were separately calculated, in which one of the In ions is replaced by Mn or Fe ions in the unit cell of the bixbyite structured In_2O_3 consisting of 80 atoms. Resulting theoretical Mn-K and Fe-K XANES spectra of Mn- and Fe-doped In_2O_3 are shown in Fig. 7 (a) and (b), respectively. As shown in this figure, characteristic features in observed XANES

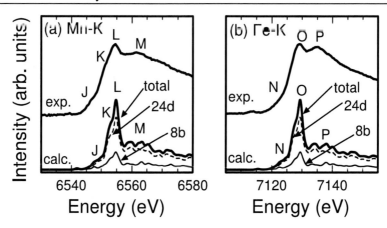

Figure 1.7. Comparison of XANES spectra of (a) Mn-K and (b) Fe-K of In_2O_3:(Mn, Fe) between experiments and calculations. Thin solid and dashed lines denote calculated spectra from Mn or Fe at 24d and 8b sites, respectively, and thick solid lines are the sums of these two. Calculated transition energies are corrected by (a)ΔE_{Mn} = -19.5 eV and (b)ΔE_{Fe}= -20.0 eV.

spectra of In_2O_3:(Mn, Fe) are well reproduced by the present models and the calculated spectral profiles are clearly distinguishable from the calculated ones for reference oxide materials. It is noted that the transition energy is also well reproduced by the present calculations, if the same amount of the energy shifts as those for the reference oxides, i.e., ΔE_{Mn} and ΔE_{Fe}, were given. From these results, it can be concluded that Mn and Fe ions are substituted at In site in the present specimen, $(In_{0.96}Mn_{0.03}Fe_{0.03})_2O_3$.

1.3.3.   Phosphor Materials

Rare-earth doped oxide materials with the perovskite type of structure have been extensively studied because of their applications in phosphor. Here Pr-doped $SrTiO_3$ and $CaTiO_3$, which are known as red light emitting phosphors [12], have been chosen and the charge state of doped Pr ions in $SrTiO_3$ and $CaTiO_3$, which plays a key role to understand the luminescence property, was examined by chemical shift in XANES spectra. Commercially available high purity powders of $SrCO_3$, $CaCO_3$, TiO_2 and Pr_2O_3 with cation ratio, i.e., (Sr+Ti):Pr and (Ca+Ti):Pr, of 1.99:0.01 were mixed and ground in an agate mortar, which were

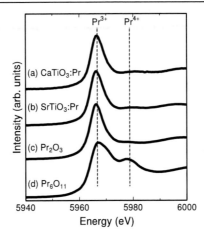

Figure 1.8. Observed Pr-L$_3$ XANES spectra of Pr-doped (a) CaTiO$_3$, (b) SrTiO$_3$ and the reference Pr oxides ((c) Pr$_2$O$_3$ and (d) Pr$_6$O$_{11}$).

calcined in air for 2 hours at 1423 K. Resulting powders were mixed and ground again, which were pressed into a pellet form. Finally, these pellets were sintered in air for 6 hours at 1523 K. In order to determine the charge state of Pr ions in SrTiO$_3$ and CaTiO$_3$, Pr-L$_3$ XANES spectra were observed, which are shown in Fig. 8 together with those of reference oxide materials, i.e., Pr$_2$O$_3$ and Pr$_6$O$_{11}$. In Pr$_2$O$_3$ and Pr$_6$O$_{11}$, charge states of Pr ions are Pr^{3+} and Pr^{3+} + Pr^{4+} with a ratio of Pr^{3+}: Pr^{4+}=1:2, respectively. As shown in Fig. 8, only one intense peak is observed in Pr-L$_3$ XANES of Pr$_2$O$_3$, which originates from Pr^{3+}, while two peaks from Pr^{3+} and Pr^{4+} are seen in that of Pr$_6$O$_{11}$. By comparing the peak energies of XANES spectra of Pr-doped SrTiO$_3$ and CaTiO$_3$ with those of the reference spectra, charge states of Pr ions in SrTiO$_3$ and CaTiO$_3$ can be determined experimentally, which show Pr ions are Pr^{3+} in these two perovskite type oxides. We also performed the first-principles calculations for Pr-doped SrTiO$_3$ and CaTiO$_3$. Prior to the calculations of Pr-doped SrTiO$_3$ and CaTiO$_3$, Pr-L$_3$ XANES of reference Pr oxide, i.e., Pr$_2$O$_3$, was examined. In this calculation, 2x2x2 supercell (40 atoms) of hexagonal structured unit cell of Pr$_2$O$_3$ was employed. Calculated Pr-L$_3$ XANES spectrum of Pr$_2$O$_3$ is compared with the experimental one in Fig. 9. Although the experimental Pr-L$_3$ XANES spectrum of Pr$_2$O$_3$ has only one sharp peak with a satellite shoulder in higher energy side, the experimental profile is well reproduced by the present calculation, when

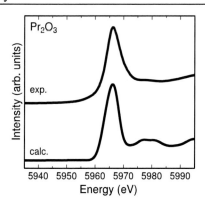

Figure 1.9. Comparison of Pr-L$_3$ XANES spectra of Pr$_2$O$_3$ between experiment and calculation.

the transition energy is corrected by -16.0 eV. For the XANES calculations of Pr-doped CaTiO$_3$ and SrTiO$_3$, one Sr^{2+} ion is replaced by Pr^{3+} ion in 2x2x2 supercell (40 atoms) of cubic perovskite structured unit cell of SrTiO$_3$ and in 2x2x1 supercell (80 atoms) of orthorhombic perovskite structured unit cell of CaTiO$_3$ with a space group of Pbnm, respectively. Calculated Pr-L$_3$ XANES spectra of Pr-doped SrTiO$_3$ and CaTiO$_3$ are compared with experimental ones in Fig. 10 (a) and (b), respectively. Only one intense peak can be seen in both of the calculated spectra of Pr-doped SrTiO$_3$ and CaTiO$_3$, which reproduced the experimental ones. It was also reported that the additional doping of Al and Ga ions in SrTiO$_3$:Pr and CaTiO$_3$:Pr increase the intensity of red light photolu-minescence [13,14]. Here the local environment of additionally doped Ga ions in SrTiO$_3$:Pr is investigated by the Ga-K XANES. Sample was prepared by the same procedure as for Pr-doped SrTiO$_3$, in which -Ga$_2$O$_3$ was additionally used for starting material with the Ga concentration of 1.5 at% to other cations (Sr, Ti and Pr ions), i.e., (Sr+Ti+Pr):Ga = 1.97:0.03. Photoluminescence spectra of SrTiO$_3$, SrTiO$_3$:Pr and Ga-doped SrTiO$_3$:Pr here synthesized are shown in Fig. 11. As shown in this figure, additional Ga doping into Pr-doped SrTiO$_3$ in-crease the photoluminescence by a factor of approximately 2.5. Observed Ga-K XANES spectra are shown in Fig. 12 with that of β-Ga$_2$O$_3$, which show clear difference between these two spectral profiles. At first, Ga-K XANES spec-trum of β-Ga$_2$O$_3$ was calculated, which is compared with the experimental one in Fig. 13. In this calculation, 1x3x2 supercell (60 atoms) of monoclinic unit

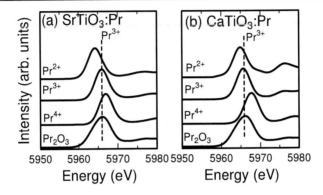

Figure 1.10. Comparison of Pr-L3 XANES spectra of Pr-doped (a) $SrTiO_3$ and (b) $CaTiO_3$ between experiments and calculations.

cell of -Ga2O3 was employed. Calculated Ga-K XANES spectrum of β-Ga_2O_3 reproduced the experimental characteristic profiles as earlier report [15], when the transition energy was corrected by ΔE = -41.5 eV. Two types of substitution models of Ga ions in $SrTiO_3$:Pr were constructed, which are the simple substitution and the vacancy models. In the former model, Sr^{2+} or Ti^{4+} are replaced by Ga^{3+} ions, while two of the Sr^{2+} or Ti^{4+} ions are replaced by two Ga^{3+} ions and one Sr^{2+} (Sr-vacancy model) or one O^{2-} (O-vacancy model) vacancies, respectively, are introduced to compensate the electronic charge of the calculated supercells in the later model. In the Sr-vacancy model, the most distant two Sr ions are replaced by two Ga ions and one nearest neighboring Sr ion from one of the substituted Ga ion is removed from the cell, while the first nearest Ti ion pair is replaced by a pair of Ga ions and O ion between these two is removed in the O-vacancy model [16]. Calculated Ga-K XANES spectra of these substituted models are compared with the experimental one in Fig. 14. Significant difference between these calculated spectra appear, in which calculated XANES spectrum of O-vacancy model shows best comparison with the observed one than those of other models. This means the Ga ions are likely to substitute at Ti site associated with oxygen vacancy in $SrTiO_3$. Another type of analysis by extended X-ray absorption fine structure (EXAFS) was also carried out. Experimental radial structure functions of Sr, Ti of $SrTiO_3$ and Ga of Ga-doped $SrTiO_3$:Pr deduced from Sr-K, Ti-K and Ga-K EXAFS spectra are shown in Fig. 15. Resultant radial structure function of Ga ion is similar to that of Ti,

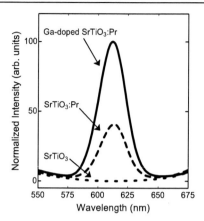

Figure 1.11. Photoluminescence spectra of Ga-doped $SrTiO_3$:Pr excited by UV (365 nm).

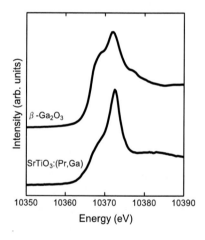

Figure 1.12. Observed Ga-K XANES spectra of Ga-doped $SrTiO_3$ and β-Ga_2O_3.

but quite different from that of Sr. This result also indicates that Ga ion must sit on the Ti site, but not on Sr site. It is sometimes dangerous to determine the local environment of doped ions by only one experimental or one theoretical evidence. However, in the present case, all of the results here investigated, i.e.,

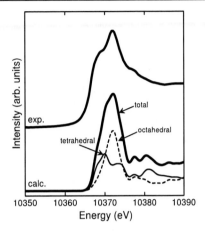

Figure 1.13. Comparison of Ga-K XANES spectra of β-Ga$_2$O$_3$ between experiment and calculation.

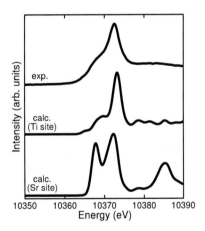

Figure 1.14. Comparison of Ga-K XANES spectra of Ga-doped SrTiO$_3$:Pr between experiment and calculations.

1) Ga-K XANES measurement, 2) theoretical XANES and 3) EXAFS, show doped Ga ions are likely to substitute at Ti site in SrTiO$_3$.

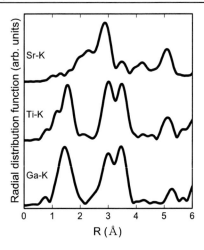

Figure 1.15. Experimental radial structure functions of Sr, Ti of $SrTiO_3$ and Ga of Ga-doped $SrTiO_3$:Pr from Sr-K, Ti-K and Ga-K EXAFS, respectively.

1.3.4.   Electrolyte of Solid Fuel Cell

It was reported that CeO_2 doped with divalent and/or trivalent cations has high oxygen ion conductivity [17]. In order to understand the mechanism of this high oxygen conductivity, it is essential to understand the vacancy formation mechanism due to the doping of divalent and/or trivalent cations. Here Y-doped CeO_2 is chosen to investigate the vacancy formation mechanism by combined use of Y-L_3 XANES and first-principles calculation. Sample specimen was prepared from the mixture of high purity powders of CeO_2 and Y_2O_3 with an atomic molar ratio of Ce:Y = 0.95:0.05, which were calcined at 1523 K for 10 hours in air after mixing and grinding, and sintered at 1773 K for 3 hours in air. Observed Y-L_3 XANES spectra of Y-doped CeO_2 and reference oxide, i.e., Y_2O_3, are shown in Fig. 16. Clear difference between these observed spectra is appeared. At first, the first-principles XANES spectrum of Y_2O_3 was obtained by using the bixbyite structured unit cell of Y_2O_3 (80atoms) to calibrate the theoretical transition energy, which is compared with experimental one in Fig. 17. Characteristic feature is well reproduced by the present calculation with the energy correction of -12.4 eV. Next the Y-L_3 XANES spectrum of Y-doped CeO_2 was investigated by using the four types of representative vacancy models, which are shown in Fig. 18. Since it is widely accepted that oxygen vacancy

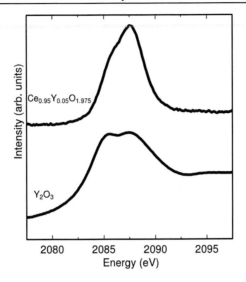

Figure 1.16. Observed Y-L$_3$ XANES spectra of Y$_2$O$_3$ and Y-doped CeO$_2$.

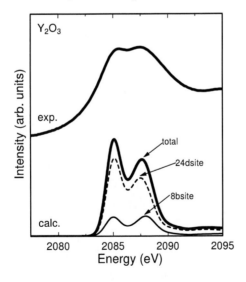

Figure 1.17. Comparison of Y-L$_3$ XANES spectra of Y$_2$O$_3$ between experiment and calculation.

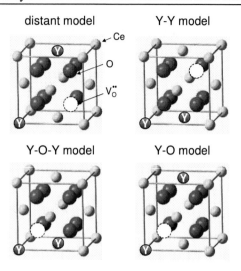

Figure 1.18. Illustration of calculated models of Y-doped CeO_2.

is introduced when Y ions are doped in CeO_2, it is assumed here that the substitution, $2Y^{3+} \rightarrow 2Ce^{4+} + O^{2-}$, occurs to control a charge balance in the cell. In the first and second models, two nearest neighboring Ce ions are replaced by two Y ions and oxygen vacancy is created between these two substituted Y ions (Y-O-Y model), and at distant position from these two Y ions (Y-Y model), respectively. In the third and fourth models, distant two Ce ions are replaced by two Y ions and oxygen vacancy is placed at the first nearest neighboring site from one of the substituted Y ions (Y-O model), and at distant site from both of the substituted Y ions (distant model), respectively. Calculated Y-L_3 XANES spectra of these four vacancy models with different configurations of substituted Y ions and oxygen vacancy are compared with experimental one for Y-doped CeO_2 in Fig. 19. Spectral width becomes broader when the oxygen vacancy is located at the first nearest neighboring site from the doped Y ions, i.e., in Y-O and Y-O-Y models, which yield the disagreement with the experimental XANES profile. Both of the calculated spectra of Y-Y and distant models reproduced the experimental one. It is noted that significant difference could not be seen when the oxygen vacancy is apart from the Y ions, which means Y-L_3 XANES is not so sensitive to determine the second nearest ion (first nearest cation) in the present case. From this comparison, it can be concluded that the

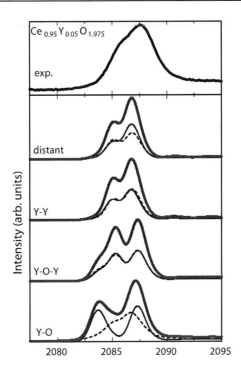

Figure 1.19. Comparison of Y-L$_3$ XANES spectra of Y-doped CeO$_2$ between experiment and calculations.

oxygen vacancy is likely to be apart from the substituted Y ions.

1.4. Conclusion

Local environment analysis of dopants in four kinds of ceramic materials, i.e., 1) bioceramics, 2) dilute magnetic semiconductor, 3) phosphor and 4) electrolyte of fuel cell, has been carried out in an atomic scale by the XANES measurements with the aid of the first-principles calculations. Observed XANES spectra have been quantitatively well reproduced by our present first-principles calculations. This type of analytical method for dopants in newly developed functional materials is very useful to understand the mechanism of their properties. In addition, it should be emphasized that there is no limitation, in principle, to apply this method to any other types of materials.

Acknowledgment

The authors would like to thank Y. Kawashima, Y. Kusakabe, S. Matsuda, Y. Mizuoka, Y. Nakade of almni of Waseda university and T. Okajima of Saga light source for their assistance in XANES measurements and first-principles calculations.

References

[1] Tanaka, I., Mizoguchi, T., Matsui, M., Yoshioka, S., Adachi, H., Ya-mamoto, T., Okajima, T., Umesaki, M., Ching, W. Y., Inoue, Y., Mizuno, M., Araki, H. and Shirai, Y. "Identification of ultra-dilute dopants in ceramics". *Nat. Mater.*, **2**, (**2003**), 541-545.

[2] Rehr, J. J. and Albers, R. C., "Theoretical approaches to x-ray absorption fine structure". *Rev. Mod. Phys.*, **72**, (2000), 621-654.

[3] Tanaka, I., Mizoguchi, T. and Yamamoto, T., "XANES and ELNES in ceramic science". *J. Am. Ceram. Soc.*, **88**, (2005), 2013-2029.

[4] Mizoguchi, T., Tanaka, I., Yoshioka, S., Kunisu, M., Yamamoto,. T,, and Ching, W. Y., "First-principles calculations of ELNES/XANES of selected wide-gap materials: Dependence on crystal structure and orientation". *Phys. Rev. B*, **70**, (2004), 045103.

[5] Mo, S. D. and Ching, W. Y., "Ab initio calculation of the core-hole effect in the electron energy-loss near-edge structure". *Phys. Rev. B*, **62**, (2000), 7901-7907.

[6] Blaha, P., Schwarz, K., Madsen, G., Kvasicka, D., and Luitz, J., "WIEN2k, An Augmented Plane Wave+Local Orbitals Program for Calculating Crystal Properties (Karlheinz Schwarz, Techn. Universitat Wien, Austria)" (2001).

[7] Monkhorst, H. J. and Pack, J. D.,"Special points for Brillouin-zone integrations". *Phys. Rev. B*, **13**, (1976), 5188-5192.

[8] Kawabata, K., Sato, H. and Yamamoto, T.,"Local environment analysis of Mn ions in β-tricalcium phosphate". *J. Ceram. Soc. Jpn.*, **116**, (2007), 108-110.

[9] Ohno, H., Munekata, H., Penny, T., von Molnar, S. and Chang, L. L., "Magnetotransport properties of p-type (In,Mn)As dilute magnetic III-V semiconductors". *Phys. Rev. Lett.*, **68**, (1992), 2664-2667.

[10] Ohno, H., Shen, A., Matsukura, F., Oiwa, A., Endo, A., Katsumoto, S. and Iye, Y., "(Ga,Mn)As: A new diluted magnetic semiconductor based on GaAs". *Appl. Phys. Lett.*, **69**, (1996), 363-365.

[11] Peleckis, G., Wang, X. L. and Dou, S. X., "Room-temperature ferromagnetism in Mn and Fe codoped In_2O_3". *Appl. Phys. Lett.*, **88**, (2006), 132507.

[12] Jia, W. Y., Xu, W. L., Rivera, I., Perez, A. and Fernandez, F., "Effects of compositional phase transitions on luminescence of $Sr_{1-x}Ca_xTiO_3 : Pr^{3+}$". *Solid State Commun.*, **126**, (2003), 153-157.

[13] Tang, J., Yu, X., Yang, L., Zhou, C. and Peng, X., "Preparation and Al^{3+} enhanced photoluminescence properties of $CaTiO_3:Pr^{3+}$". *Mat. Lett.*, **60**, (2006), 326-329.

[14] Okamoto, S. and Yamamoto, H., "Characteristic enhancement of emission from $SrTiO_3 : Pr^{3+}$ by addition of group-IIIb ions ". *Appl. Phys. Lett.*, **78**, (2001), 655-657.

[15] Okajima, T., Yamamoto, T., Kunisu, M., Yoshioka, S., Tanaka, I. and Norimasa, U., "Dilute Ga dopant in TiO_2 by X-ray absorption near-edge structure". *Jpn. J. Appl. Phys.*, **45**, (2006), 7028-.

[16] Several models for Sr- and O-vacancy models were calculated changing the configurations of positions of vacancies by the first-principles projector augmented wave package (VASP). After geometry optimization with this package, total electronic energies of the calculated models are examined, in which two models, i.e., Sr- and O-vacancy models, written in the main text have the lowest total electronic energies among the calculated models, respectively.

[17] Eguchi, K., Setoguchi, T., Inoue, T. and Arai, H., "Electrical-properties of ceria-based oxides and their application to solid oxide fuel-cells". *Solid State Ionics*, **52**, (1992), 165-172.

In: Molecular Dynamics of Nanobiostructures   ISBN: 978-1-61324-320-6
Editor: K. Kholmurodov   © 2012 Nova Science Publishers, Inc.

Chapter 2

CHROMOPHORE REARRANGEMENT IN BINDING POCKET OF RHODOPSIN MAKES SENSE FOR ITS PHYSIOLOGICAL DARK-ADAPTED STATE: COMPUTER MOLECULAR SIMULATION STUDY

*Tatyana B. Feldman[1,2], Kholmirzo T. Kholmurodov[2,3]
and Mikhail A. Ostrovsky[1,2,*]*

[1]Institute of Biochemical Physics, Russian Academy of
Sciences, Moscow, Russia
[2]JINR (Joint Institute for Nuclear Research) &
[3]Dubna International University,
Dubna, Moscow Region, Russia

Abstract

Rhodopsin protein is a typical member of G-protein-coupled receptor
(GPCR) family. It is the only receptor from a wide variety of the GPCRs

*E-mail addresses: feldman@sky.chph.ras.ru; mirzo@jinr.ru

for which a tertiary structure is known. Thus, rhodopsin represents as an excellent model to investigate the main properties and function of these receptors. In this paper the computer MD (molecular dynamics) simulation results of rhodopsin (with 11-*cis*-retinal chromophore) are presented. It is shown that 11-*cis*-retinal chromophore to be rearranged, after its insertion into the chromophore pocket. Namely, the beta-ionone ring of 11-*cis*-retinal is twisted in time frame of 0.4 ns starting from the simulation run. The changes in behavior of the nearest amino acid residues, surrounding the chromophore retinal, correlate with the beta-ionone ring twist. A clear correlation is observed between the beta-ionone ring twist and the mobility of the cytoplasmic domain that is responsible for binding of G-protein and stabilization of alpha-helix H-VI which is a characteristic site for the dark rhodopsin. The computer simulation results are discussed with the actual role that 11-*cis*-retinal plays as a chromophore group which enables rhodopsin for the ultra-fast and efficient photoisomerization and as a ligand-antagonist that keeps rhodopsin as a G-protein-coupled receptor in an inactivated state.

Keywords: MD simulations, Rhodopsin, 11-*cis*-retinal chromophore

2.1.   Introduction

The phototransduction and other biochemical cascades in living cells are mediated by a superfamily of membrane receptors known as G-protein-coupled receptors (GPCRs) [1]. All of the membrane GPCR-receptors possesses the same conformational entity, viz. the form of a seven transmembrane (TM) helical structure [2]. However, the detailed X-ray or NMR- structures of the GPCR, with the exception of the visual pigment rhodopsin, are still unknown. Rhodopsin is the first GPCR with a less or more well-defined tertiary structure [3-5], thereby representing itself as an excellent candidate for investigation of molecular details of all GPCRs family. It is worth noting that until now we cannot answer yet many important questions so that be able to testify them through the direct experimental measurement. For example, the resolution of the modern diffraction experimental setup are not sufficient enough to determine the reliable 3-dimensional structure of the chromophore group at the physiological temperature scales, on which the high-quality excited-state calculations are generally fulfilled. It is clear, that the correct chromophore conformation is necessary to achieve a more precise model of the rhodopsin reaction center, viz.

chromophore binding pocket of rhodopsin. These questions could be addressed for computer simulations and theoretical model calculations. It is known that in the dark-adapted rhodopsin the retinal chromophore is in its 11-*cis*-isomeric form and it acts as an ultra fast and an efficient molecular switch as well as a powerful antagonist that stabilizes the inactive conformation of rhodopsin (as G-protein-coupled receptor). The rhodopsin photoisomerization produces the retinal *all-trans* form that becomes as a powerful agonist, triggering thereby the transition of rhodopsin, namely metarhodopsin II, into an active state that could be able to interact with a G-protein (transducin). Followed by the photoisomerization, as the only photochemical reaction in vision, the generation of the active state (metarhodopsin II) hence is a second key event of phototransduction process.

In this work we have employed the molecular dynamics simulation method to study the behavior of the dark-adapted state of rhodopsin with 11-*cis*-retinal chromophore. The molecular dynamics simulations have been fulfilled under the physiological temperature (300K).

2.2. Conformation Behavior of 11-*cis*-Retinal: Rearrangement ("Adaptation") Process inside Chromophore Pocket

In Fig.1 a snapshot is shown for the rhodopsin protein, containing 11-*cis*-retinal chromophore, which is surrounded by a water bath. Separately in Fig.2 the 11-*cis*-retinal is presented, where the atomic and methyl group positions are shown. It is known that the 11-*cis*-chromophore retinal possesses in dark rhodopsin a twisted and a distorted conformation, unlike to that 11-*cis*-retinal possesses in solution one [6,7]. It is also doubtless that the protein binding pocket, surrounding the 11-*cis*-retinal, plays a key role in the retinal conformation and that enables chromophore retinal to exist in such of twisted and strengthen form enough stable.

In Fig.3 our simulation results on the dynamical behavior of the binding pocket (with 11-*cis*-retinal) in the beta-ionone ring region are presented. We have observed that beta-ionone ring began to rotate at the temporal point of 0.4 ns (after 400 000 steps) from the start of the simulation run. The beta-ionone ring turns with regard to the polyene chain axis to approximately 65 degrees. To estimate a "twisting degree" of different parts of chromophore retinal, we have

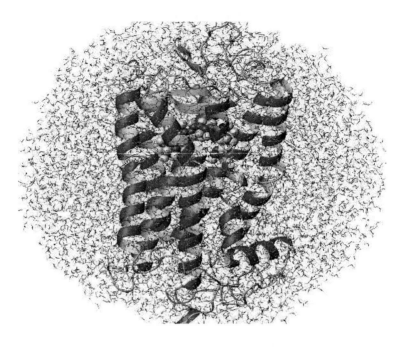

Figure 2.1. Rhodopsin protein in water solvent (PDB entry file: 1HZX). View from the side of the rhodopsin molecule. The 11-*cis*-retinal chromophore and neighbouring amino acid residue Lys296 are drawn as spherical balls.

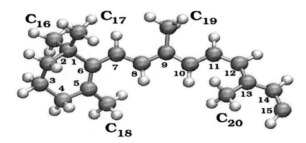

Figure 2.2. 11-*cis*-retinal chromophore: the atomic and methyl group positions.

calculated the rotational torsion angles of all methyl groups C16-C20 within the 3-ns dynamical changes. From the diagram (Fig.3, bottom diagram) it is seen that the rotation angle of the methyl group C18 is largest among others one. C18 group has seen to possess the highest amplitude of torsional oscillations. This means, by other words, that inside of rhodopsin binding pocket the C18 methyl group may serve as a "twisting key" for whole retinal polyene chain. A distorted chromophore configuration might hence to originate from the beta-ionone ring twist, resulting in rotation of the polyene chain. The retinal rearrangement process, as observed from Fig.3, we assign as a chromophore "adaptation" inside the chromophore binding pocket. Moreover, the behavior of the 11-*cis*-retinal chromophore, presented in Fig.3, are correlate well with the displacements of TM helices and amino acid residues in chromophore center and also with structural deformations of the rhodopsin cytoplasmic and extracellular domains (see below).

The details of dynamical changes we have analyzed for amino acid residues of the alpha-helix H-VI, which is located in a close neighborhood to 11-*cis*-retinal. As it is known, alpha-helix H-VI plays a stabilizing role in keeping of rhodopsin molecule in its inactive state [8]. We have found (see Fig.4) that, the distance between H-VI and 11-*cis*-retinal decreases with time. Helix H-VI approaches to the chromophore retinal by approximately 1 Å. In contrast, for the opsin model (i.e. when 11-*cis*-retinal is absent) helix H-VI has not show such of positional changes. Also, from our simulations, the residues Trp265 and Tyr268 in chromophore "adaptation" process have reach a more close position (up to about 1-2 Å) to the beta-ionone ring. It should be stressed out that this process occurs simultaneously with the beta-ionone ring twist. A minimal distance between these two amino acid residues and 11-*cis*-retinal to be about 4-5 Å, which is enough to produce a strong electrostatic interaction. In comparison to their initial states, residues Trp265 and Tyr268 at later simulation stages (from 0.4-ns to 3-ns) have strongly "surround" the 11-*cis*-retinal from two opposite sides. Their aromatic rings, as a result of the interaction with 11-*cis*-retinal chromophore, become almost parallel to each other. It is worth noting that, the simulation results are agreed with the experimental observation [8].

The behavior of Leu266 is next surprising one (Fig.5). This amino acid residue in comparison with Trp265 and Tyr268 had dramatically changed its original position; Leu266 approaches within the "adaptation" process to the chromophore retinal by approximately 7 Å. The spatial rearrangement of Leu266 is such that it will be located symmetrically against of 11-*cis*-retinal

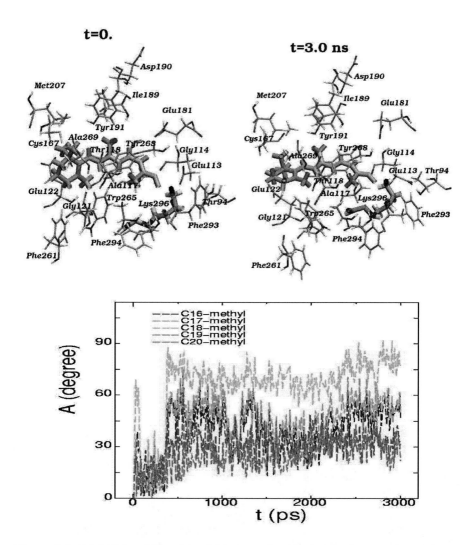

Figure 2.3. Molecular dynamics of 11-*cis*-retinal in the rhodopsin chromophore center at the initial (t=0) and final (t=3 ns) simulation states are presented along with the torsion rotation angles of five methyl groups (C16-C20) (top). The positions of the 11-*cis*-retinal atoms during the 3 ns dynamical changes are separately displayed (bottom). (View from the side of the rhodopsin molecule)

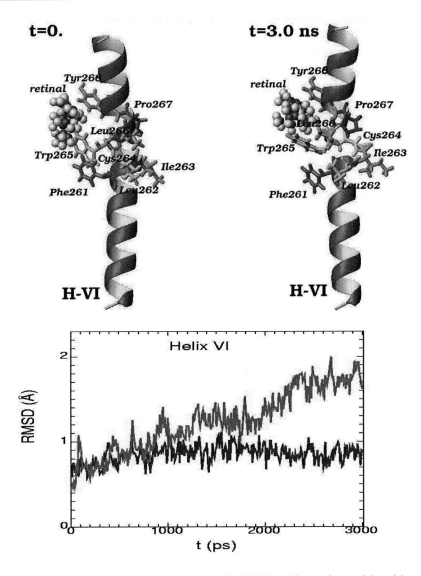

Figure 2.4. Molecular dynamics of alpha-helix H-VI and its amino acid residues surrounding the 11-*cis*-retinal chromophore are presented for the initial (t=0) and final (t=3 ns) simulation states. (The 11-*cis*-retinal is shown as balls and amino acid residues are shown as 3D structural formulas). The values of RMSD (root-mean-square-deviation) during the 3 ns molecular dynamics changes are shown for alpha-helix H-VI (bottom diagram). Blue curve - displays the deviation from the reference structure for free opsin (without retinal chromophore), red curve - for rhodopsin (with 11-*cis*-retinal chromophore).

chromophore. This is a gradually approach and this process has to be completed within a time period of about 0.4 ns from the start. We suppose that Leu266 takes a hidden part in re-orientation of the beta-ionone ring, as described in Fig.3, whereas Trp265 and Tyr268 have to act as "clamps" in this conformational rearrangement. This picture may explain the structural mechanism and valid behaviour of 11-*cis*-retinal: why the latter kept a twisted and a distorted conformation (an energetically unfavourable state) stable enough.

As for the next observation (Fig.6), within 3-ns dynamical changes and retinal rearrangement, the distance between the residues Glu122 and His211 in the rhodopsin protein moiety increases from 3 Å up to 6-7 Å. This is in contrast to that of the free opsin (without chromophore) one. This process starts at a temporal point of about 0.6 ns. One may suppose that a hydrogen bond, if existing between residues Glu122 and His211, is broken and that this happens as results of "adaptation" of 11-*cis*-retinal inside the chromophore binding pocket. The experimental data regarding the interaction peculiarities of Glu122 and His211 region are contradictory ones. On the one hand-side, the interaction of alpha-helices H-III and H-V occurs via of hydrogen bond formation among Glu122 and His211 [3,9]. On the other hand-side, in the dark-adapted rhodopsin a hydrogen bond between Glu122 and His211 may exist, but it might be broken in the rhodopsin activation process, which involves the interaction of alpha-helices H-III and H-V [10]. Thus, it is widely considered that the Glu122 and His211 interaction determines the mobility of alpha-helices H-III and H-V.

Our calculation results indicate that in dark-adapted rhodopsin its 11-*cis*-retinal separates from each other alpha-helices H-III and H-V, thereby keeping them at their distorted sites. Thus, it is worth making an assumption that the chromophore retinal photoisomerization will change the electrostatic interaction of the whole protein surrounding and that it will destroy a "distorted balance" noted above. As a result, alpha-helices H-III and H-V, are yet being fixed at their location sites, begun to relax to a state which is proper for free opsin one.

2.3.   A Rhodopsin's "Coupling Place" with Transducin

In this section we investigate the dynamical behavior of far rhodopsin domains (viz., cytoplasmic and extracellular) and in correlation with the chromophore "adaptation" process described above. The 11-*cis*-retinal "adaptation" process as discribed above one presents via Fig.7, where conformational changes at the initial (t=0) and final (t=3 ns) states are compared. The comparison shows the

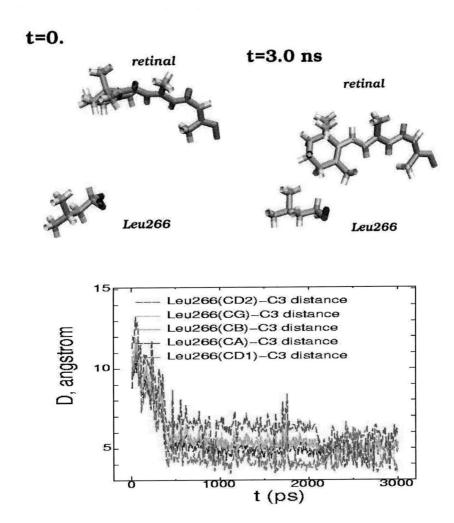

Figure 2.5. Molecular dynamics of 11-*cis*-retinal chromophore and amino acid residue Leu266 of rhodopsin are presented for the initial (t=0) and final (t=3 ns) simulation states. (Side view from the rhodopsin cytoplasmic part). The plots of the interatomic distances between the different atoms of Leu266 and atom C3 of the beta-ionone ring of 11-*cis*-retinal are displayed for the 3 ns molecular dynamics changes (bottom diagram).

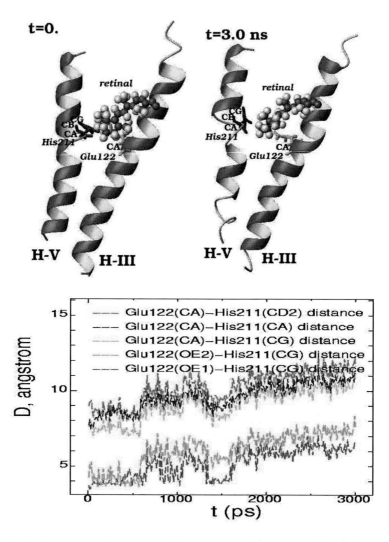

Figure 2.6. Molecular dynamics of the alpha-helices H-III (Glu122) and H-V (His211) for the initial (t=0) and final (t=3 ns) simulation states are presented for the rhodopsin with 11-*cis*-retinal chromophore. (11-*cis*-retinal is shown as balls and amino acids are shown as 3D structural formulas). The plot of the interatomic distances between the different atoms in Glu122 and His211 are displayed for the 3 ns molecular dynamics changes (bottom diagram).

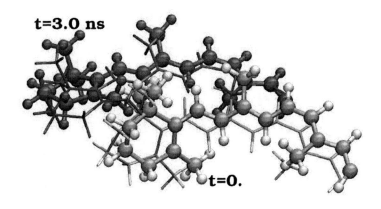

Figure 2.7. Molecular dynamics of 11-*cis*-retinal chromophore at the initial (t=0) and final (t=3 ns) simulation states are presented along five methyl groups (C16-C20). (View from the side of the rhodopsin molecule).

positions of the five methyl groups (C16-C20) inside of the rhodopsin molecule.

The 11-*cis*-retinal "adaptation" in chromophore binding center initiates the structural rearrangement not only for the nearest amino acids residues, but it also causes the essential changes in far protein domains. It is seen from Fig.8 that during of 11-*cis*-retinal "adaptation" two protein loops, viz. C-end peptide (via the amino acid residue Ser334) and C-II loop (via the amino acid residue Ala241), have reach within a short time each other and they form a very close contact. At the beginning of simulations these two residues (viz. Ser334 and Ala241) are positioned enough far away of each other. Nevertheless, after about 0.4 ns the distance between residues Ser334 and Ala241 sharply decrease from 8-10 Å to about 3 Å. In other words, starting from this temporal point of 0.4 ns a new hydrogen bond, as results of C-end peptide and C-II loop interaction, has form between of Ser334 and Ala241. It is important to note that these two residues are located in so-called "transducin coupling place" [11,12]. Such a hydrogen bond formation between Ser334 and Ala241 is not observed, however, if 11-*cis*-retinal chromophore is absent (viz. for the opsin model, rhodopsin without chromophore group) [28,29].

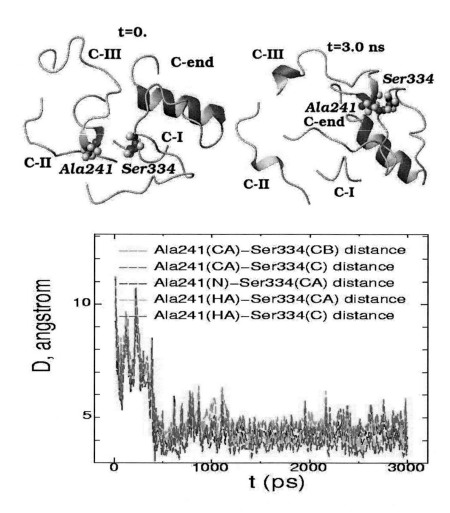

Figure 2.8.   Molecular dynamics of C-end peptide (Ser334) and C-II loop (Ala241) for the initial (t=0) and final (t=3 ns) simulation states in rhodopsin with 11-*cis*-retinal chromophore group. The plot of interatomic distances between the different atoms in Ser334 and Ala241 are displayed for the 3 ns molecular dynamics changes (bottom diagram).

2.4. Summary

To summarize, we suppose that the chromophore "adaptation" in binding pocket brought the rhodopsin molecule not only to a state of "high alert", but it also stabilizes the rhodopsin inactive conformation as a G-protein-coupled receptor. The chromophore "adaptation" process one may also consider as a mechanism that ensures for the rhodopsin protein a proper conformational shape not only in chromophore binding center but also in the far cytoplasmic or extracellular domains, preventing thereby a rhodopsin coupling with transducin as much as possible. This means, by other words, that 11-*cis*-retinal chromophore acts as a powerful antagonist that will block for the rhodopsin molecule a "coupling place" with transducin.

Methods

The molecular mechanical (MM) and molecular dynamics (MD) simulations were performed on the rhodopsin protein with a water solvent. A molecular model was constructed from the rhodopsin crystal structure (PDB: 1HZX) [3,7]. The missing amino acids 236-240 and 331-333 were inserted using the MOE software package for biomolecular simulations [13]. The MM and MD simulations were performed with the software package AMBER for the simulation of biomolecules [14,15], optimized for a special-purpose computer MDGRAPE-2 [16-18,26,27]. For the MDGRAPE-2 all the particle interactions are calculated. The all-atom force field of Cornell et al. [19,20] was used in the MD simulations. A system was solvated with TIP3P molecules [21] generated in a spherical (non-periodic) water bath. The temperature was kept constant by using Berendsen algorithm with a coupling time of 0.2 ps [22]. Only bond lengths involving hydrogen atoms were constrained using the SHAKE method [23]. The integration time step in the MD simulations was 1 fs. The simulation procedures were the same in all calculations. Firstly, a potential energy minimization was performed on an initial state for each system. Next, the MD simulation was performed on the energy-minimized states. The temperatures of the considered systems were gradually increased by heating to 300 K for and then kept at 300 K for the next 3 million time steps. The trajectories at 300 K for 3.0 ns were compared and studied in detail. The result of simulations and images of simulated proteins were analyzed by using the RasMol [24] and MOLMOL [25] packages.

Acknowledgments

The MD simulations have been performed using computing facilities, software and clusters at CICC (JINR, Russia), RICC (RIKEN, Japan), RIKEN-Yokohama (MDGRAPE-2,3), Yasuoka Laboratory (Keio University, Japan).

References

[1] Mirzadegan, T., Benko, G., Filipek, S., Palczewski K., "Sequence analyses of G-protein-coupled receptors: similarities to rhodopsin". *Biochemistry*, **42**, (2003), 2759-2767.

[2] Gether, U., Kobilka, B.K., "G protein-coupled receptor". *J. Biol. Chem.*, **273**, (1998), 17979-17982.

[3] Palczewski, K., Kumasaka, T., Hori, T., Behnke, C.A., Motoshima, H., Fox, B.A., Le Trong, I., Teller, D.C., Okada, T., Stenkamp, R.E., Yamamoto, M., Miyano M., "Crystal structure of rhodopsin: a G protein-coupled receptor". *Science*, **289**, (2000), 739-745.

[4] Liang, Y., Fotiadis, D., Filipek, S., Saperstein, D.A., Palczewski, K., Engel, A., "Organization of the G protein-coupled receptors rhodopsin and opsin in native membranes". *J. Biol. Chem.*, **278**, (2003), 21655-21662.

[5] Okada, T., Sugihara, M., Bondar, A.-N., Elstner, M., Entel, P., Buss, V., "The retinal conformation and its environment in rhodopsin in light of a new 2.2 Å crystal structure". *J. Mol. Biol.*, **342**, (2004), 571-583.

[6] Salgado, G.F.J., Struts, A.V., Tanaka, K., Fujioka, N., Nakanishi, K., Brown, M.F., "Deuterium NMR Structure of Retinal in the Ground State of Rhodopsin". *Biochemistry*, **43**, (2004), 12819-12828.

[7] Teller, D.C., Okada, T., Behnke, C.A., Palczewski, K., Stenkamp R.E., "Advances in determination of a high-resolution three-dimensional structure of rhodopsin, a model of G-protein-coupled receptor (GPCRs)". *Biochemistry*, **40**, (2001), 7761-7772.

[8] Menon, S.T., Han, M., Sakmar, T.P., "Rhodopsin: structural basis of molecular Physiology". *Physiol. Rev.*, **81**, (2001), 1659-1688.

[9] Beck, M., Sakmar, T.P., Siebert, F., "Spectroscopic evidence for interaction between transmembrane helices 3 and 5' rhodopsin". *Biochemistry*, **37**, (1998), 7630-7639.

[10] Patel, A.B., Crocker, E., Eilers, M., Hirshfeld, A., Sheves, M., Smith, S.O., "Coupling of retinal isomerization to the activation of rhodopsin". *Proc. Natl. Acad. Sci. USA.*, **101**, (2004), 10048-10053.

[11] Phillips, W.J., Cerione, R.A., "A C-terminal peptide of bovine rhodopsin binds to the transducin alpha-subunit and facilitates its activation". *Biochem. J.*, **299**, (1994), 351-357.

[12] Yamashita, T., Terakita, A., Shichida, Y., "Distinct roles of the second and third cytoplasmic loops of bovine rhodopsin in G protein activation". *J. Biol. Chem.*, **275**, (2000), 34272-34279.

[13] MOE (Molecular Operating Environment) (http://www.chemcomp.com; used within 2002-2003, by license of CAL RIKEN).

[14] Pearlman, D.A., Case, D.A., Caldwell, J.W., Ross, W.R., Cheatham, T.E., DeBolt, S., Ferguson, D., Seibel, G., Kollman, P., "AMBER, a computer program for applying molecular mechanics, normal mode analysis, molecular dynamics and free energy calculations to elucidate the structures and energies of molecules". *Comp. Phys. Commun.*, **91**, (1995), 1-41.

[15] Case, D.A., Pearlman, D.A., Caldwell, J.W., Cheatham, T.E., Ross, W.S., Simmerling, C.L., Darden, T.A., Merz, K.M., Stanton, R.V., Cheng, A. L., Vincent, J.J., Crowley, M., Ferguson, D.M., Radmer, R.J., Seibel, G.L., Singh, U.C., Weiner, P.K., Kollman P.A., "AMBER 8.0. University of California". (2003).

[16] Narumi, T., Susukita, R., Ebisuzaki, T., McNiven, G., Elmegreen B., "Molecular Dynamics Machine: Special-purpose Computer for Molecular Dynamics Simulations". *Molecular Simulation*, **21**, (1999), 401-408.

[17] Narumi, T., Susukita, R., Furusawa, H., Ebisuzaki, T., "46 Tflops Special-purpose Computer for Molecular Dynamics Simulations: (WINE-2)". *Proc. 5th Int. Conf. on Signal Processing. Beijing.*, (2000), 575-582.

[18] Okimoto, N., Yamanaka, K., Suenaga, A., Hirano, Y., Futatsugi, N., Narumi, T., Yasuoka, K., Susukita, R., Koishi, T., Furusawa, H., Kawai, A., Hata, M., Hoshino, T., Ebisuzaki T., "Molecular dynamics simulations of prion proteins - effect of Ala117Val mutation". *Chem-Bio Informatics J.*, **3**, (2003), 1-11.

[19] Ponder, J.W., Case, D.A., "Force fields for protein simulations". *Adv. Prot. Chem.*, **66**, (2003), 27-85.

[20] Cornell, W.D., Cieplak, P., Bayly, C.I., Gould, I.R., Merz, Jr.K.M., Ferguson, D.M., Spellmeyer, D.C., Fox, T., Caldwell, J.W., Kollman, P.A., "A second Generation forth field for the simulation of Proteins and Nucleic Acids". *J. Am. Chem. Soc.*, **117**, (1995), 5179-5197.

[21] Jorgensen, W.L., Chandrasekhar, J., Madura, J.D., "Comparison of simple potential functions for simulating liquid water". *J. Chem. Phys.*, **79**, (1983), 926-935.

[22] Berendsen, H.J.C., Postma, J.P.M., van Gunsteren, W.F., DiNola, A., Haak, J.R., "Molecular dynamics with coupling to an external bath". *J. Chem. Phys.*, **81**, (1984), 3684-3690.

[23] Ryckaert, J.P., Ciccotti, G., Berendsen, H.J.C., "Numerical integration of the Cartesian equations of proteins and nucleic acids". *J. Comput. Phys.*, **23**, (1997), 327-341.

[24] Sayle, R.A., Milner-White, E.J., "RasMol: Biomolecular graphics for all". *Trends in Biochem. Sci.*, **20**, (1995), 374-376.

[25] Koradi, R., Billeter, M., Wuthrich, K., "MOLMOL: a program for display and analysis of macromolecular structure". *J. Mol. Graphics*, **4**, (1996), 51-55.

[26] Kholmirzo Kholmurodov (Ed.), "Molecular Simulation Studies in Material and Biological Sciences", *Nova Science Publishers Ltd.*, (2007), 190p., ISBN 1-59454-607-x.

[27] Kholmurodov, K.T., Hirano, Y., Ebisuzaki, T., "MD Simulations on the Influence of Disease-Related Amino Acid Mutations in the Human Prion Protein". *Chem-Bio Informatics Journal*, **3**, No. 2, (2003), 86.

[28] Kh. T. Kholmurodov, Kh.T., Feldman, T.B., Ostrovskii, M.A., "Molecular dynamics of rhodopsin and free opsin: Computer simulation". *Neuroscience and Behavioral Physiology*, **37**, (2007), 161.

[29] Kholmurodov, Kh.T., Feldman, T.B., Ostrovsky, M.A., "Interaction of chromophore, 11-*cis*-retinal, with amino acid residues of the visual pigment rhodopsin in the region of protonated Schiff base: a molecular dynamics study". *Russian Chemical Bulletin, International Edition*, **56**, No. 1, (2007), 20.

In: Molecular Dynamics of Nanobiostructures ISBN: 978-1-61324-320-6
Editor: K. Kholmurodov © 2012 Nova Science Publishers, Inc.

Chapter 3

MOLECULAR DYNAMICS STUDY OF HUMAN RED BLOOD CELL MEMBRANE

Aram A. Shahinyan[1], Paruyr K. Hakobyan[2]
and Armen H. Poghosyan[2,*]
[1]Institute of Applied Problems of Physics,
National Academy of Sciences,
Republic of Armenia
[2]International Scientific-Educational Center
National Academy of Sciences,
Republic of Armenia

Abstract

We have performed a 80 ns molecular dynamics (MD) simulation of
an asymmetric model of the human red blood cell (erythrocyte) membrane and have investigated its dynamical and structural properties. The
system consists of all major phospholipids found in the human erythrocyte membrane in a real experiment, cholesterol molecules, and the transmembrane part of Glycophorin A (GpA) – an important intrinsic protein

*E-mail address: artsha@sci.am

of the erythrocyte membrane. The surrounding molecular composition of the protein is investigated and the influence of surrounding phospholipid and cholesterol molecules on protein properties is discussed. The NAMD code with the corresponding CHARMM27 force field was used. Some structural parameters of the membrane, such as the area per molecule, membrane thickness, surface roughness, etc are discussed.

Keywords: Molecular dynamics, Human blood cells, Erythrocyte membrane

3.1.   Introduction

The biological membrane is a major component of a living cell. Over the last years, a lot of real and virtual experiments have been done on biological membranes [1,2,3,4,5]. Currently, research on biological membranes is widely reported. The molecular dynamics simulation (MD) method [4,5,6,7], which is actively used in studying biosystems, is a great tool to understand well the intra- and intermolecular structure of biological systems [8,9,10]. Recently, a lot of MD-based research has been done on the structure and behavior of biological membranes [10–14].

The main purpose of this work is the modeling and investigation of human erythrocyte membrane and comparison of some parameters with results of real experiments, as well as studying the GpA protein transmembrane part and the influence of the surrounding molecular composition (phospholipids, cholesterol, and water molecules) on the structural properties of the protein.

The effect of boundary neighboring lipids and cholesterol on protein is still a debated issue, and to the best of our knowledge, there has been no computer simulation study of multi-component complex systems with an embedded protein. A few works were done on Dipalmitoylphosphatidylcholine (DPPC) and Dimyristoylphosphatidylcholine (DMPC) with embedded GpA pure systems; also, GpA has been intensively studied in the presence of Sodium dodecylsulfate micelle [13] and in vacuum [15].

An asymmetric and more accurate model of the human red blood erythrocyte membrane leads to a real simulation of the biological membrane and will help us to understand the behavior of the membrane and protein along with its surrounding phospholipids.

3.2.   Construction and Simulation Details

The model membrane was constructed with the use of Hyperchem (Hyper-cube Inc.) software and MDesigner [17,18] based on experimental data on the molecular composition of the asymmetric membrane of the human erythrocyte (Table 1).

For the construction of the human erytrocite membrane, the molecules of Phosphadityl ethanolamine (POPE), Phosphadityl choline (POPC), Phos-phadityl ethanolamine (SAPE), Phosphadityl choline (SOPC), Phosphadityl serine (SAPS), Phosphadityl serine (SDPS), Sphingomyelin (LSM), Sphin-gomyelin (HSM), and Cholesterol were created, and we received a system of 252 molecules of phospholipids, cholesterol, and 27 Sodium counterions – by random replication, making allowance for the asymmetry of the model mem-brane and the final concentration of phospholipids and cholesterol. The trans-membrane part of the GpA protein was inserted into the system that had been developed in the previous simulation. The investigated model was solvated by insertion into water bulk with 8572 water molecules of the TIP3P [19] model. The membrane model was hydrated at about 33 water molecules per phospho-lipid in order to assure that the system is fully hydrated. The membrane model size was about 10.5 x 9 x 9 nm^3, so the initial estimation of the area per phos-pholipid was about 0.89 nm^2. The initial configuration of the membrane was determined by an energy minimization using the conjugate gradient method for 15000 steps and was subjected to a short (about 1000 ps) MD simulation in a NVT ensemble. The Langevin dynamics [20] with a damping coefficient of $5ps^{-1}$ was used. The constant temperature and pressure were set to 310K and normal 1 atm, respectively. The constant temperature and pressure were con-trolled by using the Langevin piston Nose-Hoover method [21]. For the non-bonded full electrostatic interactions between atom pairs, the cutoff parameter was set to 14 Å. The visual representation was performed with a VMD pack-age. A MD simulation run (in the NPT ensemble) was done with a timestep of 2 fs, using NAMD software code with the CHARMM27 all-atom force field on a parallel Linux cluster. The simulation was performed for 80 ns on 12 nodes (24 Intel Xeon 3.06 GHz processors) of the cluster during about 60 days. The force field parameters of the cholesterol molecule were generated using a Dundee PRODRG server. The system ready for MD simulations consisted of 57640 atoms.

3.3.   Results and Discussion

As is mentioned above, a model of the phospholipid bilayer of the asymmetric erythrocyte membrane has been constructed based on experimental data on the molecular composition of the human erythrocyte membrane (see Table 1). The table shows that the molecules of the erythrocyte membrane phospholipids differ in the length of the hydrocarbon chains and in their saturation degree.

Figure 1 shows changes in the free energy of the asymmetric phospholipid bilayer of the human erythrocyte membrane during a simulation using the molecular dynamics method.

Table 3.1. Proportion of phospholipids, cholesterol, and sphingomyelin in the model of the phospholipid bilayer of the asymmetric erythrocyte membrane

Name	Acyl Chain	Concentr. in outer layer (%)	Concentr. in inner layer (%)	Avg Consistence, (%)	Chem. Symb.
sphingomyelin	24:0–14:0	19.8	2.3	21.8	LSM
sphingomyelin	16:0-14:0	20.6	2.3		HSM
phosphadityl serine	18:0-20:4w6	0	9.9	10.7	SAPS
phosphadityl serine	18:0-22:6w3	0	10.8		SDPS
phosphaditylethanol amine	16:0 – 18:1	1.6	18.3	21.4	POPE
phosphaditylethanol amine	18:0 – 20:4	3.3	18.3		SAPE
phosphadityl choline	16:0 – 18:1	14.9	7.6	22.6	POPC
phosphadityl choline	18:0 – 18:1	15.7	7.6		SOPC
cholesterol	—	24.1	22.9	23.5	Chol

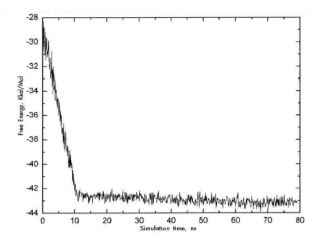

Figure 3.1. Dependence of the free energy of the asymmetric phospholipid bilayer of erythrocytes on MD simulation time.

The estimations of the free energy of biological membranes [23] obtained by experimental methods show that it is in the range 89.7±35.4 kcal/mol. By comparing these data with results of our computer experiment (Fig. 1), we can argue that after 12-14 ns of simulation we received an erythrocyte membrane model in the thermodynamic equilibrium state with the free energy of -43±0.5 kcal/mol. Figure 2 represents the model of the human erythrocyte membrane after 14 ns MD simulation

Since the asymmetric erythrocyte membrane model is in equilibrium, it becomes topical to study the structural parameters and dynamical behavior of the system. One of the most important structural parameters of the phospholipid bilayer received from a physics experiment is the interlayer spacing (the phospholipid bilayer thickness). Fig. 3 shows the variation of the thickness of the phospholipid bilayer of an asymmetric erythrocyte membrane depending on the MD simulation time. As can be seen in Figure 3, the membrane thickness grows up to 60 ns of simulation and then remains practically constant at 51.5±0.5 Å. This result is comparable with the experimentally determined thickness of the erythrocyte membrane: 55±0.5 Å [16]. On the other hand, the time-averaged thickness of the phospholipid bilayer can be estimated from the electron density profile of the system. The peak values obtained for the electron density make it

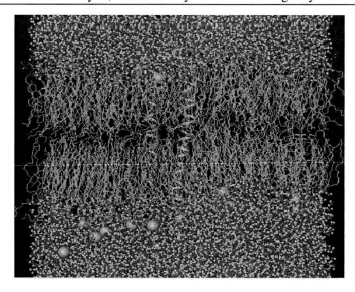

Figure 3.2. Human erythrocyte membrane model after 14 ns MD simulation.

possible to calculate the relative positions of the head groups of the phospholipid molecules.

Figure 4 represents the electron density of the asymmetric erythrocyte membrane model in the direction of the bilayer normal (along the z axis). From Figure 4, the thickness of the membrane can also be obtained, which corresponds to the distance between the peaks of the two layers.

It should be noted that the obtained data (about 52 Å) are quite comparable with the value of the membrane thickness in Figure 3 after 60 ns of MD simulation. Another main structural parameter of the biological membranes is the average area per molecule. For individual phospholipids (SAPS, SDPS, POPE, SAPE, LSM, HSM, POPC, and SOPC) in asymmetric erythrocyte membranes, it was calculated by the Voronoi algorithm. Due to the asymmetry of the erythrocyte membrane, it would be more correct to represent separately the values of the average area per molecule for the outer and inner layers of the membrane. Figure 5 (a,b,c,d) represents the average area per molecule for phospholipids of the asymmetric membrane of an erythrocyte depending on the MD simulation time. Each figure displays changes in the average area for both the outer and inner layers of the membrane.

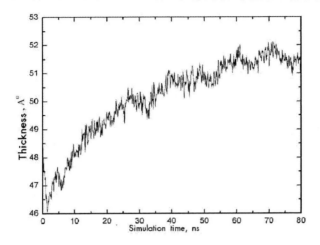

Figure 3.3. The thickness of the asymmetric erythrocyte membrane depending on the simulation time.

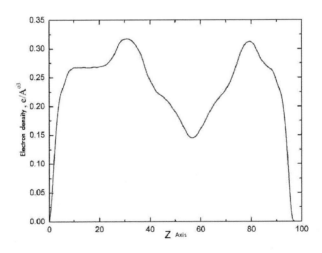

Figure 3.4. The electron density of the asymmetric erythrocyte membrane along the z axis.

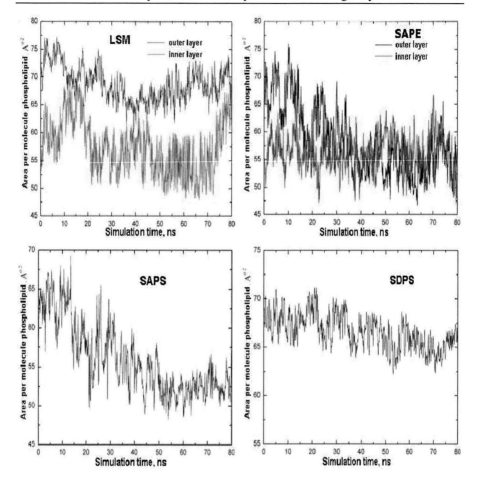

Figure 3.5. Area per molecule for phospholipids depending on the MD simulation time. (a) LSM, (b) SAPE, (c) SAPS, (d) SDPS.

As is seen in the figures, the area per molecule of LSM is about 67 Å^2 and 57 Å^2; for HSM, it is 65 Å^2 and 55 Å^2 for the outer and inner layers of the membrane, respectively. The difference between the values for these phospholipid molecules of the outer and inner layers can be explained by the asymmetry of the system, where the phospholipid composition of the outer and inner layers of the membrane is different. The average area per molecule of sphingomyelin throughout the erythrocyte membrane is about 60 Å². At the same time, it is known that the average area per molecule in pure sphingomyelin bilayers is about 52 Å² [24]. The results are comparable with experimental data. Due to the absence of the phosphaditylserine molecules on the outer side of the plasma membrane, in constructing a model of asymmetric erythrocyte membrane, the molecules were only presented in the inner layer of the system. Figure 5 (c, d) shows that the mean area per phosphaditylserine molecules (SDPS and SAPS) decreases over the simulation time and becomes almost constant after 60 ns of MD simulation. The values of the average area for SDPS and SAPS molecules are about 68 Å² and 53 Å², respectively. The average area per phosphaditylserine molecule across the erythrocyte membrane is about 61 Å². The obtained MD data are in agreement with results (about 58 Å²) received from symmetric systems consisting of three types of phospholipids (PC, PE, and PS).

To visualize the erythrocyte membrane surface, we investigated the surface roughness of the phospholipid bilayer. The surface roughness is calculated as follows:

where $z(r)$ are the z coordinates of the two phosphorus atoms of polar groups of phospholipid molecules on the surfaces of the bilayer. Figure 6 shows a plot of this parameter for both sides of the bilayer.

It can be seen from the figure that the values of surface roughness for both layers of the phospholipid bilayer of the asymmetric erythrocyte membrane reach on the average up to 5.5 Å.

In MD and X-ray analysis of the erythrocyte membrane, the average coefficient of surface roughness of the membrane varies in the range 0.65 ± 0.1 nm [25].

For a better understanding of the general state of the system, it is interesting to study the behavior of cholesterol molecules in the membrane. In this regard, we have calculated the few parameters which describe the behavior and localization of cholesterol molecules. The density of cholesterol molecules in the erythrocyte asymmetric membrane was determined along the normal to the bilayer surface, i.e., the z axis (Figure 8). The figure shows that the cholesterol molecules are mainly concentrated in the hydrophobic part of the membrane – more precisely, under the polar heads of phospholipids molecules. It is known that cholesterol molecules are the seals of the membrane and mostly play a constructive role. It can be assumed (Figure 8) that being located mainly in the hydrophobic part of the membrane, where there are the hydrocarbon chains of phospholipid molecules, cholesterol is responsible for the so-called "close pack-

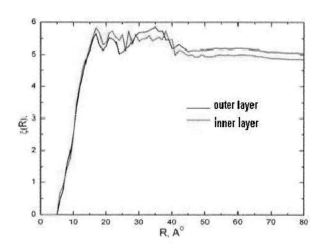

Figure 3.6. Surface roughness function of erythrocyte asymmetric membrane model.

Figure 3.7. Snapshot of phospholipids phosphorus atom ordering after 80 ns of MD simulation.

ing" of the membrane (adding cholesterol molecules leads to a closer packing of the hydrocarbon chains of phospholipid molecules).

In the asymmetric erythrocyte membrane model, the transmembrane part of the glycophorin A protein has been introduced and its basic properties have been studied. One of the main structural parameters characterizing the protein is the angle between the A and B helices. In this paper, we investigated the dynamics of the change in the angle between the A and B helices of glycophorin in the asymmetric membrane of an erythrocyte. The angle between the A and B helices of glycophorin was calculated as follows. At each step of the simulation, each helix was considered to be a vector directed from the center of mass of the last amino acid to the center of mass of the first amino acid of the helix, and the angle between the helices was determined as the angle between these vectors. Figure 9 shows the change in the angle between the A and B helices of glycophorin depending on the MD simulation time. As can be seen from the figure, the tilt angle between the A and B helices varies in the range of $33^O \pm 2$ to $42^O \pm 2$, reaching the maximum at about 53^0; after 60 ns of simulation, it is almost constant. From NMR studies, it was found that an increase in the degree of the unsaturation of the hydrocarbon chains in the mono-phospholipid bilayer leads to an increase in the angle between the helices of the glycophorin dimer [8]. With increasing the length of the hydrocarbon chains of the phospholipid molecules from 14 to 20 CH_2 groups, the angle difference between the helices becomes 12^0.

In our case, when instead of mono phospholipid bilayer, we investigate an asymmetric multi-component phospholipid bilayer containing a molecule of cholesterol and phospholipids with hydrocarbon chains with a length of 16-24 CH_2 groups, the angle between the helices of the glycophorin molecule is on the average about 40°. It follows that a more detailed study of protein conformation in an erythrocyte asymmetric membrane is needed to investigate the behavior of molecules of phospholipids located in the neighboring media of the protein. In order to clarify the nature of the conformational changes of glycophorin in the erythrocyte membrane, the dynamic behavior of molecules of phospholipids and cholesterol in the neighboring media of the protein has been studied. The investigations were carried out as follows: the protein was virtually placed in a cylinder with a radius of 10 Å, and phospholipid molecules and cholesterol were examined in a virtual cylinder with a radius of 20 Å with the same center. This means that we consider the molecules of phospholipids and cholesterol in a circle with a radius of 10 Å from the glycophorin (Fig. 10). To

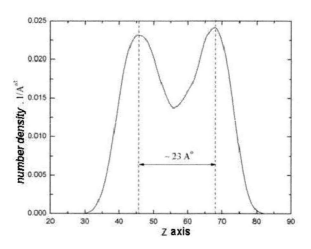

Figure 3.8. Density of cholesterol molecules in the asymmetric membrane of an erythrocyte along the normal to the bilayer surface.

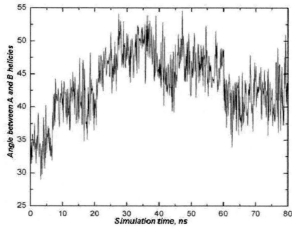

Figure 3.9. Angle between the A and B helices of glycophorin in the asymmetric membrane of an erythrocyte depending on the MD simulation time.

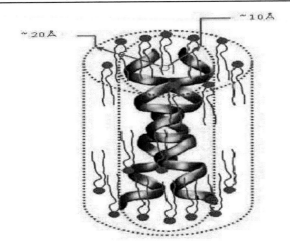

Figure 3.10. Surroundings of glycophorin A (GpA).

determine the phospholipid composition in the surrounding of glycophorin (at a distance of 10 Å), a change in the concentration of phospholipid molecules and cholesterol at the beginning and the end of a MD simulation were investigated. These data were compared with the corresponding concentrations throughout the membrane. The data are presented in Table 2 (a) and (b). Since the membrane is asymmetric as regards the content of phospholipids, the data for the outer and inner layers of the bilayer differ from each other.

The tables show that in the simulation, there is both the accumulation and removal of phospholipids and cholesterol in the environment of the protein. For example, there is more than a twofold increase in the concentration of phosphatidylserine around the protein in the inner half of the membrane, whereas the same phospholipid is completely absent in the other half of the membrane.

There was a significant decrease in the concentration of cholesterol in the space around glycophorin. Thus, glycophorin has in its immediate vicinity the following concentrations of phospholipids: phosphatidylserine (40%), phosphatidylethanolamine (about 33%), sphingomyelin (about 17%), and cholesterol (15%). The tables show that in the inner layer of the membrane, phosphatidylserine molecules surround GpA, and in the outer layer of the membrane sphigomyelin, phosphatidylcholine and cholesterol molecules are present.

Based on the results, of particular importance is the detailed dynamic struc-

Table 3.2. Concentration (%) of phospholipid molecules and cholesterol in the immediate surrounding of GpA at the beginning of the simulation

Phospholipids	Chol	PC (POPC+ SOPC)	PE (POPE+ SAPE)	PS (SDPS+ SAPS)	SPH (LSM+ HSM)
Inner layer (protein surrounding)	14.4	9.5	28.5	47.6	0
Inner layer (average concentration in the membrane)	22.9	15.1	36.6	20.6	4.8
Outer layer (protein surrounding)	24.8	25.2	10.6	0	39.4
Outer layer (average concentration in the membrane)	24.1	30.6	4.8	0	40.5

ture of glycophorin in the erythrocyte membrane. For clarity, we have designated the distance between the center of masses (COM) of amino acids as follows: 0 to 10 Å is dark gray; 10- 15 Å is gray; 15 -20 Å is light gray; and more than 20 Å is white. The strongly contacting amino acids are localized in the so-called "first shell" (dark gray). From the starting point, one can notice a lot of close or strong contacts.

The observed inter-helix distances are the results of intermolecular interactions and are generally governed by the immediate neighboring (helix-dipole and helix-chain) lipid environments. The starting point configuration of the GpA protein molecule was extracted from our previously simulated system with the overall simulation time of about 100 ns (a PC/PE/GpA/water system). However, after additional 80 ns of simulation, some "close contacts" disappeared, which means that the distance between amino acids became more than 10 Å. In figure 11(a), the strongly connected amino acids from helices A and B are shown.

It is known from NMR experiments that Gly79 plays an important role in GpA dimerization, and hydrogen bonds between Gly79, Val80, and Gly83 of two helixes stabilize the dimer [13]. We estimated the distances of the above mentioned amino acids and represented the distance changes during the whole simulation time. In addition, we examined the contact distance between Leu75 and Ile76 in order to explain the effect of this contact in dimerization. The

**Table 3.3. Concentration (%) of phospholipids and cholesterol in the
immediate surrounding of GpA at the end point of the simulation**

Phospholipids	Chol	PC (POPC+ SOPC)	PE (POPE+ SAPE)	PS (SDPS+ SAPS)	SPH (LSM+ HSM)
Inner layer (protein surrounding)	13	13	21.7	47.8	4.5
Inner layer (average concentration in the membrane)	22.9	15.1	36.6	20.6	4.8
Outer layer (protein surrounding)	20	30.5	10	0	40
Outer layer (average concentration in the membrane)	24.1	30.6	4.8	0	40.5

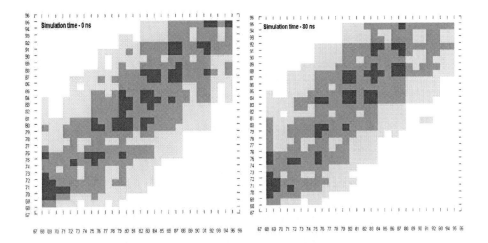

Figure 3.11. Contact matrix of amino acid residues (distance between the centers of mass) between helices A and B of GpA (horizontal axis: helix A; vertical axis: helix B) at the beginning (a) and end (b) of simulation. The distances between the amino acid residue pairs are represented by squares: dark gray for 0–10 Å, gray for 11–15 Å, light gray for 16–20 Å, and white for over 20 Å.

Gly79 – Val80 distance fluctuates in the range of 5 Å⊥1 and 7 Å±1 during the whole simulation time. Indeed, we can see a strong and close contact over the whole simulation time. In comparison with experimental data (NMR refinements), we can see a shift (about 3 Å) of the Gly79 – Val80 distance. The NMR value of the Gly79 – Val80 distance is 3.5 Å, which is almost the starting point of our simulation (Fig. 11(b)). During 80 ns of MD simulation, one can see an increase in the distance of up to 7 Å and equilibrium as well after 60 ns of simulation. The increase is due to the presence of a surrounding medium – namely, phospholipids and cholesterol molecules. Changing the medium – for instance, introducing surfactants (SDS micelle) [9] – leads to a decrease in the Gly79 – Val80 distance to the value of 3.5 Å, which means that the surrounding medium influences the GpA structural parameters. For an explanation, we also visualized and presented the snapshots extracted from the last point of simulation (at 80 ns). We also measured the hydrogen – hydrogen distance. From the structural parameters, we found that the equilibrium state is reached after 60 ns of simulation, and therefore, the Gly79 – Val80 distance is estimated to be about 7 Å. The hydrogen – hydrogen distance is about 4.5 Å as shown in Fig. 12.

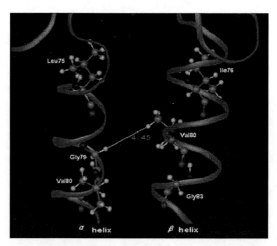

Figure 3.12. GpA after a 80 ns simulation. The amino acid residues Gly79, Val80, Gly80, Gly83, Leu75, and Ile76, as well as the location and distance between the hydrogen atoms of the amino acid residues Gly79 and Val80 in helices A and B are noted.

3.4. Conclusion

By means of a computer experiment, the most realistic model of a human ery-throcyte membrane was studied. Important data were received which are not yet accessible in a physical experiment. An equilibrium model of the human erythrocyte asymmetric membrane was constructed, and a comparison of ex-perimental findings shows that the calculated structural parameters of the mem-brane are in good agreement with the data obtained in real experiments. The main feature of this work is a detailed study and discussion of the GpA protein and its structural and conformational parameters. The conformational changes of the GpA protein in the asymmetric membrane of an erythrocyte are mainly related to the surrounding phospholipid and cholesterol molecules interacting with the protein molecule, and the unsaturation degree and length of the acyl chains of the phospholipid molecules. It is also noted that the Gly 79 – Val 80 amino acid residues are responsible for the dimerization of the GpA protein.

References

[1] Yeghiazaryan G.A., Poghosyan A.H., Shahinyan A.A., "The water molecules orientation around the dipalmitoylphosphatidylcholine head group: A molecular dynamics study". *Physica A*, **362**, (2006), 197-203.

[2] Tieleman D.P., Berendsen H.J.C., "Molecular dynamics simulations of a fully hydrated dipalmitoylphosphatidylcholine bilayer with different macroscopic boundary conditions and parameters". *J. Chem. Phys.*, **105**, (1996), 4871-4880.

[3] Chiu S.W., Clark M., Balaji V., Subramaniam S., Scott L.H., Jakobsson E., "Incorporation of surface tension into molecular dynamics simulation of an interface: a fluid phase lipid bilayer membrane". *Biophys. J.*, **69**, (1995), 1230-1245.

[4] Poghosyan A.H., Gharabekyan H.H., Shahinyan A.A., "Molecular Dynamics Simulation of DMPC/DPPC mixed bilayers". *IJMPC*, **18**, (2007), 73 - 89.

[5] Mori K., Hata M., Neya S., Hoshino T., "MD simulation of asymmetric phospholipid bilayers with ions and cholesterol". *Chem-Bio Inf. J.*, **4**, (2004), 15-26.

[6] Robinson A.J., Richards W.G., Thomas P. J., Hann M.M., "Behvior of chlesterol and it's effect on head group and chain conformations in lipid bilayers: a molecular dynamics study". *Biophys. J.*, **67**, (1994), 2345-2354.

[7] Berkowitz M.L., "Detailed molecular dynamics simulations of model biological membranes containing cholesterol". *Biochimica et biophysica acta. Biomembranes*, **86**, (2009), 86-96.

[8] White S.H., Wimley W.C., "Membrane Protein folding and stability: Phys ical principles". *Annu. Rev. Biophys. Biomol. Struct.*, **28**, (1999), 319-366.

[9] Braun R., Engelman D.M., Schulten K., "Molecular Dynamics Simulation of Micelle Formation around dimeric glycophorin A transmembrane helices". *Biophys. J.*, **87**, (2004), 754-763.

[10] Soumana O.S., Garnier N., Genest M., "Molecular Dynamics simulation for the prediction of transmembrane helix-helix heterodimers assembly". *Eur. Biophys. J.*, **36**, (2007), 1071-1082.

[11] Zhang J., Lazaridis T., "Transmembrane Helix Association Affinity Can Be Modulated by Flanking and Noninterfacial Residues". *Biophys. J.*, **96**, (2009), 4418-4427.

[12] Eilers M., Patel A.B., Liu W., Smith S.O., "Comparison of Helix Interactions in Membrane and Soluble a-Bundle Proteins". *Biophys. J.*, **82**, (2002), 2720-2736.

[13] Smith S.O., Song D., Shekar S., Groesbeek M., Ziliox M., Aimoto S., "Structure of the transmembrane dimer interface of glycophorin A in membrane bilayers". *Biochemistry*, **40**, (2001), 6553-6558.

[14] Smith S.O., Eilers M., Song D., Crocker E., Ying W., Groesbeek M., Metz G., Zilox M., Aimoto S., "Implications of Threonine hydrogen bonding in the glycophorin A transmembrane helix dimer". *Biophys. J.*, **82**, (2002), 2476-2486.

[15] Beevers A.J., Kukol A., "Systematic molecular dynamics searching in a lipid bilayer: application to the glycophorin A and oncogenic Erb-2 transmembrane domains". *J. of Mol. Graphics and Modelling*, **25**, (2006), 226-233.

[16] Sackmann E., "Biological Membranes. Architecture and Function". *Handbook of Biological Physics. 1, edited by R. Lipowsky and E. Sackmann.*, **(1995)**.

[17] Hypercube Inc. URL: http://www.hyper.com/.

[18] Shahinyan A.A., Poghosyan A.H., Yeghiazaryan G.A., Gharabekyan H.H., *Elec J. Nat. Sci.*, **1**, (2004), 56-61.

[19] Jorgensen W.L., Chandrasekhar J., Medura J.D., Impey R.W., Klein M.L., "Comparison of Simple Potential Functions for Simulating Liquid Water". *J. Chem. Phys.*, **79**, (1983), 926-935.

[20] Levy R.M., McCammon J.A., Karplus M., "Diffusive Langevin dynamics of model alkanes" *Chem. Phys. Lett*, **64**, (1979), 4-11.

[21] Feller S.E., Zeng Y.H., Pastor R.W., Brooks B.R., "Constant pressure molecular dynamics simulation: The Langevin piston method". *J. Comp. Phys.*, **103**, (1995), 4613-4621.

[22] Petrache H.I., Grossfield A., MacKenzie K.R., Engelman D.M., Woolf T.B., "Modulation of Glycophorin A transmembrane helix interactions by lipid bilayers: Molecular Dynamics calculations". *J. Mol. Biol.*, **302**, (2000), 727-746.

[23] Sandermann H., "High free energy of lipid/protein interaction in biological membranes". *FEBS Letters*, **514**, (2002), 340-342.

[24] Hyvonen M.T., Kovanen P.T., "Molecular Dynamics Simulation of sphingomyelin bilayers". *J. Phys. Chem. B*, **107**, (2003), 9102-9108.

[25] Schmidt C.F., Svoboda K., Lei N., Petsche I.B., Berman L.E.,Safinya C.R., Grest G.S., "Existence of a flat phase in red cell membrane skeleton". *Science*, **259**, (1993), 952-955.

In: Molecular Dynamics of Nanobiostructures   ISBN: 978-1-61324-320-6
Editor: K. Kholmurodov   © 2012 Nova Science Publishers, Inc.

Chapter 4

GENERALIZED-ENSEMBLE SIMULATIONS IN PROTEIN SCIENCE

*Yoshiharu Mori[a], Ayori Mitsutake[b]
and Yuko Okamoto[a,c,*]*
[a]Department of Physics, Nagoya University
Nagoya, Aichi 464-8602, Japan
[b]Department of Physics, Keio University
Yokohama, Kanagawa 223-8522, Japan
[c]Structural Biology Research Center
Nagoya University, Nagoya, Aichi 464-8602, Japan

Abstract

In simulations in materials and biological sciences, one encounters with a great difficulty that conventional simulations will tend to get trapped in states of energy local minima. A simulation in generalized ensemble performs a random walk in potential energy space and can overcome this difficulty. From only one simulation run, one can obtain canonical-ensemble averages of physical quantities as functions of temperature by the histogram reweighting techniques. We review the

[*]E-mail addresses: ymori@tb.phys.nagoya-u.ac.jp; ayori@mail.rk.phys.keio.ac.jp;
okamoto@phys.nagoya-u.ac.jp

generalized-ensemble algorithms. The multidimensional extensions of the replica-exchange method and simulated tempering are presented. The effectiveness of the methods is tested with short peptide and protein systems.

Keywords: Monte Carlo; Molecular dynamics; Generalized-ensemble algorithm; Replica-exchange method; Simulated tempering

4.1. Introduction

Conventional Monte Carlo (MC) and molecular dynamics (MD) simulations in materials and biological sciences are greatly hampered by the multiple-minima problem. The canonical fixed-temperature simulations at low temperatures tend to get trapped in a few of a huge number of local-minimum-energy states which are separated by high energy barriers. One way to overcome this multiple-minima problem is to perform a simulated annealing (SA) simulation [1], and it has been widely used in biomolecular systems (see, e.g., Refs. [2]– [8] for earlier applications). The SA simulation mimics the crystal-making process, and temperature is lowered very slowly from a sufficiently high temperature to a low one during the SA simulation. As the temperature is lowered, the Boltzmann weight factor is accordingly changed, and so the thermal equilibrium is continuously broken. Hence, accurate thermodynamic averages for fixed temperatures cannot be obtained.

A class of simulation methods, which are referred to as the *generalized-ensemble algorithms*, overcome both above difficulties, namely the multipole-minima problem and inaccurate thermodynamic averages (for reviews see, e.g., Refs. [9]– [15]). In the generalized-ensemble algorithm, each state is weighted by an artificial, non-Boltzmann probability weight factor so that a random walk in potential energy space may be realized. The random walk allows the simulation to escape from any energy barrier and to sample much wider conformational space than by conventional methods. Unlike SA simulations, the weight factors are fixed during the simulations so that the eventual reach to the thermal equilibrium is guaranteed. From a single simulation run, one can obtain accurate ensemble averages as functions of temperature by the single-histogram [16] and/or multiple-histogram [17, 18] reweighting techniques (an extension of the multiple-histogram method is also referred to as the *weighted histogram analysis method* (WHAM) [18]).

One of effective generalized ensemble algorithms is the *replica-exchange method* (REM) [19]. (The method is also referred to as *parallel tempering* [20]). In this method, a number of non-interacting copies (or, replicas) of the original system at different temperatures are simulated independently and simultaneously by the conventional MC or MD method. Every few steps, pairs of replicas are exchanged with a specified transition probability. The details of molecular dynamics algorithm for REM, which is referred to as the *Replica-Exchange Molecular Dynamics* (REMD) have been worked out in Ref. [21], and this led to a wide application of REM in the protein science [10, 15].

Another effective generalized-ensemble algorithm is *simulated tempering* (ST) [22,23] (the method is also referred to as the *method of expanded ensemble* [22]) performs a free random walk in temperature space. This random walk, in turn, induces a random walk in potential energy space and allows the simulation to escape from states of energy local minima.

One is naturally led to a multidimensional (or, multivariable) extension of REM, which we refer to as the *multidimensional replica-exhcange method* (MREM) [24]. (The method is also referred to as *Hamiltonian replica-exchange method* [25].) General formulations for multidimensional generalized-ensemble algorithms have also been worked out [26–28], and special versions for isobaric-isothermal ensemble have been developed [29, 30].

In this Chapter, we describe the generalized-ensemble algorithms mentioned above. Namely, we review the multidimensional REM and ST [24, 26–28, 30]. The effectiveness of these methods are tested with short peptide and protein systems.

4.2. Multidimensional Generalized-Ensemble Algorithms

4.2.1. General Formulations

We now give the general formulations for the multidimensional generalized-ensemble algorithms [26–28]. Let us consider a generalized potential energy function $E_\lambda(x)$, which depends on L parameters $\lambda = (\lambda^{(1)}, \cdots, \lambda^{(L)})$, of a system in state x. Although $E_\lambda(x)$ can be any function of λ, we consider the following specific generalized potential energy function for simplicity:

$$E_\lambda(x) = E_0(x) + \sum_{\ell=1}^{L} \lambda^{(\ell)} V_\ell(x) . \tag{4.2.1}$$

Here, there are $L+1$ energy terms, $E_0(x)$ and $V_\ell(x)$ ($\ell = 1, \cdots, L$), and $\lambda^{(\ell)}$ are the corresponding coupling constants for $V_\ell(x)$.

After integrating out the momentum degrees of freedom, the partition function of the system at fixed temperature T and parameters λ is given by

$$Z(T,\lambda) = \int dx \exp(-\beta E_\lambda(x)) = \int dE_0 dV_1 \cdots dV_L \, n(E_0, V_1, \cdots, V_L) \exp\left(-\beta E_\lambda\right) ,$$
$$\tag{4.2.2}$$

where $n(E_0, V_1, \cdots, V_L)$ is the multidimensional density of states:

$$n(E_0, V_1, \cdots, V_L) = \int dx \delta(E_0(x) - E_0) \delta(V_1(x) - V_1) \cdots \delta(V_L(x) - V_L) . \tag{4.2.3}$$

Here, the integration is replaced by a summation when x is discrete.

The expression in Eq. (4.2.1) is often used in simulations. For instance, in simulations of spin systems, $E_0(x)$ and $V_1(x)$ (here, $L = 1$ and $x = \{S_1, S_2, \cdots\}$ stand for spins) can be respectively considered as the zero-field term and the magnetization term coupled with the external field $\lambda^{(1)}$. (For Ising model, $E_0 = -J\sum_{<i,j>} S_i S_j$, $V_1 = -\sum_i S_i$, and $\lambda^{(1)} = h$, i.e., external magnetic field.) In umbrella sampling [31] in molecular simulations, $E_0(x)$ and $V_\ell(x)$ can be taken as the original potential energy and the (biasing) umbrella potential energy, respectively, with the coupling parameter $\lambda^{(\ell)}$ (here, $x = \{q_1, \cdots, q_N\}$ where q_k is the coordinate vector of the k-th particle and N is the total number of particles). For the molecular simulations in the isobaric-isothermal ensemble, $E_0(x)$ and $V_1(x)$ (here, $L = 1$) correspond respectively to the potential energy U and the volume $\mathcal{V}$ coupled with the pressure $\mathcal{P}$. (Namely, we have $x = \{q_1, \cdots, q_N, \mathcal{V}\}$, $E_0 = U$, $V_1 = \mathcal{V}$, and $\lambda^{(1)} = \mathcal{P}$, i.e., E_λ is the enthalpy without the kinetic energy contributions.) For simulations in the grand canonical ensemble with N particles, we have $x = \{q_1, \cdots, q_N, N\}$, and $E_0(x)$ and $V_1(x)$ (here, $L = 1$) correspond respectively to the potential energy U and the total number of particles N coupled with the chemical potential μ. (Namely, we have $E_0 = U$, $V_1 = N$, and $\lambda^{(1)} = -\mu$.)

Moreover, going beyond the well-known ensembles discussed above, we can introduce any physical quantity of interest (or its function) as the additional potential energy term V_ℓ. For instance, V_ℓ can be an overlap with a reference configuration in spin glass systems, an end-to-end distance, a radius of gyration in molecular systems, etc. In such a case, we have to carefully choose the range of $\lambda^{(\ell)}$ values so that the new energy term $\lambda^{(\ell)}V_\ell$ will have roughly the same order of magnitude as the original energy term E_0. We want to perform a simulation where a random walk not only in the E_0 space but also in the V_ℓ space is realized. As shown below, this can be done by performing a multidimensional REM, ST, or MUCA simulation.

We first describe the *multidimensional replica-exchange method* (MREM) [24]. The crucial observation that led to this algorithm is: As long as we have M *non-interacting* replicas of the original system, the Hamiltonian $H(q,p)$ of the system does not have to be identical among the replicas and it can depend on a parameter with different parameter values for different replicas. The system for the multidimensional REM consists of M non-interacting replicas of the original system in the "canonical ensemble"' with $M(= M_0 \times M_1 \times \cdots \times M_L)$ different parameter sets Λ_m ($m = 1, \cdots, M$), where $\Lambda_m \equiv (T_{m_0}, \lambda_m) \equiv (T_{m_0}, \lambda_{m_1}^{(1)}, \cdots, \lambda_{m_L}^{(L)})$ with $m_0 = 1, \cdots, M_0, m_\ell = 1, \cdots, M_\ell$ ($\ell = 1, \cdots, L$). Because the replicas are non-interacting, the weight factor is given by the product of Boltzmann-like factors for each replica:

$$W_{\text{MREM}} \equiv \prod_{m_0=1}^{M_0} \prod_{m_1=1}^{M_1} \cdots \prod_{m_L=1}^{M_L} \exp\left(-\beta_{m_0} E_{\lambda_m}\right). \qquad (4.2.4)$$

Without loss of generality we can order the parameters so that $T_1 < T_2 < \cdots < T_{M_0}$ and $\lambda_1^{(\ell)} < \lambda_2^{(\ell)} < \cdots < \lambda_{M_\ell}^{(\ell)}$ (for each $\ell = 1, \cdots, L$). The multidimensional REM is realized by alternately performing the following two steps:

1. For each replica, a "canonical" MC or MD simulation at the fixed parameter set is carried out simultaneously and independently for a certain steps.

2. We exchange a pair of replicas i and j which are at the parameter sets Λ_m and Λ_{m+1}, respectively. The transition probability for this replica exchange process is given by

$$w(\Lambda_m \leftrightarrow \Lambda_{m+1}) = \min\left(1, \exp(-\Delta)\right), \qquad (4.2.5)$$

where we have

$$\Delta = (\beta_{m_0} - \beta_{m_0+1}) \left(E_{\lambda_m}\left(q^{[j]}\right) - E_{\lambda_m}\left(q^{[i]}\right)\right), \qquad (4.2.6)$$

for T-exchange, and

$$\Delta = \beta_{m_0} \left[\left(E_{\lambda_{m_\ell+1}}(q^{[j]}) - E_{\lambda_{m_\ell+1}}(q^{[i]}) \right) - \left(E_{\lambda_{m_\ell}}(q^{[j]}) - E_{\lambda_{m_\ell}}(q^{[i]}) \right) \right],$$
(4.2.7)

for $\lambda^{(\ell)}$-exchange (for one of $\ell = 1, \cdots, L$). Here, $q^{[i]}$ and $q^{[j]}$ stand for configuration variables for replicas i and j, respectively, before the replica exchange.

We now consider the *multidimensional simulated tempering* (MST) which realizes a random walk both in temperature T and in parameters λ [26–28]. The entire parameter set $\Lambda = (T, \lambda) \equiv (T, \lambda^{(1)}, \cdots, \lambda^{(L)})$ become dynamical variables and both the configuration and the parameter set are updated during the simulation with a weight factor:

$$W_{\mathrm{MST}}(\Lambda) \equiv \exp\left(-\beta E_\lambda + f(\Lambda)\right),$$
(4.2.8)

where the function $f(\Lambda) = f(T, \lambda)$ is chosen so that the probability distribution of Λ is flat:

$$P_{\mathrm{MST}}(\Lambda) \propto \int dE_0 dV_1 \cdots dV_L \, n(E_0, V_1, \cdots, V_L) \, \exp\left(-\beta E_\lambda + f(\Lambda)\right) \equiv \text{constant}.$$
(4.2.9)

This means that $f(\Lambda)$ is the dimensionless ("Helmholtz") free energy:

$$\exp\left(-f(\Lambda)\right) = \int dE_0 dV_1 \cdots dV_L \, n(E_0, V_1, \cdots, V_L) \, \exp(-\beta E_\lambda).$$
(4.2.10)

In the numerical work we discretize the parameter set Λ in $M(= M_0 \times M_1 \times \cdots \times M_L)$ different values: $\Lambda_m \equiv (T_{m_0}, \lambda_m) \equiv (T_{m_0}, \lambda_{m_1}^{(1)}, \cdots, \lambda_{m_L}^{(L)})$, where $m_0 = 1, \cdots, M_0, m_\ell = 1, \cdots, M_\ell$ ($\ell = 1, \cdots, L$). Without loss of generality we can order the parameters so that $T_1 < T_2 < \cdots < T_{M_0}$ and $\lambda_1^{(\ell)} < \lambda_2^{(\ell)} < \cdots < \lambda_{M_\ell}^{(\ell)}$ (for each $\ell = 1, \cdots, L$). The free energy $f(\Lambda_m)$ is now written as $f_{m_0, m_1, \cdots, m_L} = f(T_{m_0}, \lambda_{m_1}^{(1)}, \cdots, \lambda_{m_L}^{(L)})$.

Once the initial configuration and the initial parameter set are chosen, the multidimensional ST is realized by alternately performing the following two steps:

1. A "canonical" MC or MD simulation at the fixed parameter set $\Lambda_m = (T_{m_0}, \lambda_m) = (T_{m_0}, \lambda_{m_1}^{(1)}, \cdots, \lambda_{m_L}^{(L)})$ is carried out for a certain steps with the weight factor $\exp(-\beta_{m_0} E_{\lambda_m})$ (for fixed Λ_m, $f(\Lambda_m)$ in Eq. (4.2.8) is a constant and does not contribute).

2. We update the parameter set Λ_m to a new parameter set $\Lambda_{m\pm1}$ in which one of the parameters in Λ_m is changed to a neighboring value with the configuration and the other parameters fixed. The transition probability of this parameter updating process is given by the following Metropolis criterion:

$$w(\Lambda_m \to \Lambda_{m\pm1}) = \min\left(1, \frac{W_{\mathrm{MST}}(\Lambda_{m\pm1})}{W_{\mathrm{MST}}(\Lambda_m)}\right) = \min\left(1, \exp\left(-\Delta\right)\right) .$$

$$(4.2.11)$$

Here, there are two possibilities for $\Lambda_{m\pm1}$, namely, T-update and $\lambda^{(\ell)}$-update. For T-update, we have $\Lambda_{m\pm1} = (T_{m_0\pm1}, \lambda_m)$ with

$$\Delta = (\beta_{m_0\pm1} - \beta_{m_0}) E_{\lambda_m} - (f_{m_0\pm1,m_1,\cdots,m_L} - f_{m_0,m_1,\cdots,m_L}) . \qquad (4.2.12)$$

For $\lambda^{(\ell)}$-update (for one of $\ell = 1, \cdots, L$), we have $\Lambda_{m\pm1} = (T_{m_0}, \lambda_{m_\ell\pm1})$ with

$$\Delta = \beta_{m_0}(E_{\lambda_{m_\ell\pm1}} - E_{\lambda_{m_\ell}}) - (f_{m_0,\cdots,m_\ell\pm1,\cdots} - f_{m_0,\cdots,m_\ell,\cdots}) , \qquad (4.2.13)$$

where
$$\lambda_{m_\ell\pm1} = (\cdots, \lambda_{m_{\ell-1}}^{(\ell-1)}, \lambda_{m_\ell\pm1}^{(\ell)}, \lambda_{m_{\ell+1}}^{(\ell+1)}, \cdots) \quad \text{and} \quad \lambda_{m_\ell} = (\cdots, \lambda_{m_{\ell-1}}^{(\ell-1)}, \lambda_{m_\ell}^{(\ell)}, \lambda_{m_{\ell+1}}^{(\ell+1)}, \cdots).$$

4.2.2. Weight Factor Determinations for the Multidimensional ST

Among the three multidimensional generalized-ensemble algorithms described above, only MREM can be performed without much preparation because the weight factor for MREM is just a product of regular Boltzmann-like factors. On the other hand, we do not know the MST weight factors *a priori* and need to estimate them. As a simple method for these weight factor determinations, we can generalize the *replica-exchange simulated tempering* [32] to multidimensions.

Suppose we have made a single run of a short MREM simulation with $M(= M_0 \times M_1 \times \cdots \times M_L)$ replicas that correspond to M different parameter sets Λ_m $(m = 1, \cdots, M)$. Let $N_{m_0,m_1,\cdots,m_L}(E_0, V_1, \cdots, V_L)$ and $n_{m_0,m_1,\cdots,m_L}$ be respectively the $(L+1)$-dimensional potential-energy histogram and the total number of samples obtained for the m-th parameter set $\Lambda_m = (T_{m_0}, \lambda_{m_1}^{(1)}, \cdots, \lambda_{m_L}^{(L)})$. The generalized WHAM equations are then given by

$$
n(E_0, V_1, \cdots, V_L) = \frac{\displaystyle\sum_{m_0, m_1, \cdots, m_L} N_{m_0, m_1, \cdots, m_L}(E_0, V_1, \cdots, V_L)}{\displaystyle\sum_{m_0, m_1, \cdots, m_L} n_{m_0, m_1, \cdots, m_L} \exp\left(f_{m_0, m_1, \cdots, m_L} - \beta_{m_0} E_{\lambda_m}\right)} ,
$$

(4.2.14)

and

$$
\exp(-f_{m_0, m_1, \cdots, m_L}) = \sum_{E_0, V_1, \cdots, V_L} n(E_0, V_1, \cdots, V_L) \exp\left(-\beta_{m_0} E_{\lambda_m}\right). \quad (4.2.15)
$$

The density of states $n(E_0, V_1, \cdots, V_L)$ and the dimensionless free energy $f_{m_0, m_1, \cdots, m_L}$ (which is the MST parameter) are obtained by solving Eqs. (4.2.14) and (4.2.15) self-consistently by iteration.

4.2.3. Expectation Values of Physical Quantities

We now present the equations to calculate ensemble averages of physical quantities with any temperature T and any parameter λ values.

After a long production run of MREM and MST simulations, the canonical expectation value of a physical quantity A with the parameter values Λ_m ($m = 1, \cdots, M$), where $\Lambda_m \equiv (T_{m_0}, \lambda_m) \equiv (T_{m_0}, \lambda_{m_1}^{(1)}, \cdots, \lambda_{m_L}^{(L)})$ with $m_0 = 1, \cdots, M_0, m_\ell = 1, \cdots, M_\ell$ ($\ell = 1, \cdots, L$), and $M(= M_0 \times M_1 \times \cdots \times M_L)$, can be calculated by the usual arithmetic mean:

$$
<A>_{T_{m_0}, \lambda_m} = \frac{1}{n_m} \sum_{k=1}^{n_m} A(x_m(k)) , \quad (4.2.16)
$$

where $x_m(k)$ ($k = 1, \cdots, n_m$) are the configurations obtained with the parameter values Λ_m ($m = 1, \cdots, M$), and n_m is the total number of measurements made with these parameter values. The expectation values of A at any intermediate T ($= 1/k_B\beta$) and any λ can also be obtained from

$$
<A>_{T, \lambda} = \frac{\displaystyle\sum_{E_0, V_1, \cdots, V_L} A(E_0, V_1, \cdots, V_L) n(E_0, V_1, \cdots, V_L) \exp\left(-\beta E_\lambda\right)}{\displaystyle\sum_{E_0, V_1, \cdots, V_L} n(E_0, V_1, \cdots, V_L) \exp\left(-\beta E_\lambda\right)} ,
$$

(4.2.17)

where the density of states $n(E_0, V_1, \cdots, V_L)$ is obtained from the multiple histogram reweighting techniques. Namely, from the MREM or MST simulation, we first obtain the histogram $N_{m_0, m_1, \cdots, m_L}(E_0, V_1, \cdots, V_L)$ and the total number of samples $n_{m_0, m_1, \cdots, m_L}$ in Eq. (4.2.14). The density of states $n(E_0, V_1, \cdots, V_L)$ and the dimensionless free energy $f_{m_0, m_1, \cdots, m_L}$ are then obtained by solving Eqs. (4.2.14) and (4.2.15) self-consistently by iteration. Substituting the obtained density of states $n(E_0, V_1, \cdots, V_L)$ into Eq. (4.2.17), one can calculate the ensemble average of the physical quantity A at any T and any λ.

4.3. Simulations Results

We tested the effectiveness of the generalized-ensemble algorithms by using a system of a 17-residue fragment of ribonuclease T_1 [33, 34]. It is known by experiments that this peptide fragment forms α-helical conformations [33]. We have performed a two-dimensional REM simulation and a two-dimensional ST simulation. In these simulations, we used the following energy function:

$$E_\lambda = E_0 + \lambda E_{\text{SOL}} , \qquad (4.3.18)$$

where we set $L = 1, V_1 = E_{\text{SOL}}$, and $\lambda^{(1)} = \lambda$ in Eq. (4.2.1). Here, E_0 is the potential energy of the solute and E_{SOL} is the solvation free energy. The parameters in the conformational energy as well as the molecular geometry were taken from ECEPP/2 [35–37].

The solvation term E_{SOL} is given by the sum of terms that are proportional to the solvent-accessible surface area of heavy atoms of the solute [38]. For the calculations of solvent-accessible surface area, we used the computer code NSOL [39].

The computer code KONF90 [7, 8] was modified in order to accommodate the generalized-ensemble algorithms. The simulations were started from randomly generated conformations. We prepared eight temperatures ($M_0 = 8$) which are distributed exponentially between $T_1 = 300$ K and $T_{M_0} = 700$ K (i.e., 300.00, 338.60, 382.17, 431.36, 486.85, 549.49, 620.20, and 700.00 K) and four equally-spaced λ values ($M_1 = 4$) ranging from 0 to 1 (i.e., $\lambda_1 = 0$, $\lambda_2 = 1/3$, $\lambda_3 = 2/3$, and $\lambda_4 = 1$) in the two-dimensional REM simulation and the two-dimensional ST simulation. Simulations with $\lambda = 0$ (i.e., $E_\lambda = E_0$) and with $\lambda = 1$ (i.e., $E_\lambda = E_0 + E_{\text{SOL}}$) correspond to those in gas phase and in aqueous solution, respectively.

We first present the results of the two dimensional REM simulation. We used 32 replicas with the eight temperature values and the four λ values given above. Before taking the data, we made the two-dimensional REM simulation of 100000 MC sweeps with each replica for thermalization. We then performed the two-dimensional REM simulation of 1000000 MC sweeps for each replica to determine the weight factor for the two-dimensional ST simulation. At every 20 MC sweeps, either T-exchange or λ-exchange was tried (the choice of T or λ was made randomly). In each case, either set of pairs of replicas $((1,2),...,(M-1,M))$ or $((2,3),...,(M,1))$ was also chosen randomly, where M is M_0 and M_1 for T-exchange and λ-exchange, respectively.

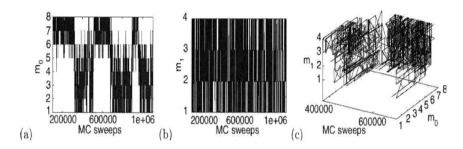

Figure 4.1. Time series of the labels of T_{m_0}, m_0, (a) and λ_{m_1}, m_1, (b) as functions of MC sweeps, and that of both m_0 and m_1 for the region from 400000 MC sweeps to 700000 MC sweeps (c). The results were from one of the replicas (Replica 1). In (a) and (b), MC sweeps start at 100000 and end at 1100000 because the first 100000 sweeps have been removed from the consideration for thermalization purpose.

In Fig. 1 we show the time series of labels of T_{m_0} (i.e., m_0) and λ_{m_1} (i.e., m_1) for one of the replicas. The replica realized a random walk not only in temperature space but also in λ space. The behavior of T and λ for other replicas was also similar (see Ref. [28]). From Fig. 1, one finds that the λ-random walk is more frequent than the T-random walk.

We now use the results of the two-dimensional REM simulation to determine the weight factors for the two-dimensional ST simulation by the multiple-histogram reweighting techniques. Namely, by solving the generalized WHAM equations in Eqs. (4.2.14) and (4.2.15) with the obtained histograms at the 32 conditions, we obtained 32 values of the ST parameters f_{m_0,m_1} ($m_0 = 1, \cdots, 8; m_1 = 1, \cdots, 4$).

After obtaining the ST weight factor, $W_{ST} = \exp(-\beta_{m_0}(E_C + \lambda_{m_1}E_{SOL}) + f_{m_0,m_1})$, we carried out the two-dimensional ST simulation of 1000000 MC sweeps for data collection after 100000 MC sweeps for thermalization. At every 20 MC sweeps, either T_{m_0} or λ_{m_1} was respectively updated to $T_{m_0 \pm 1}$ or $\lambda_{m_1 \pm 1}$ (the choice of T or λ update and the choice of ± 1 were made randomly).

We show the average total energy, average conformational energy, average $\lambda \times E_{SOL}$, and average end-to-end distance in Fig. 2. The results are in good agreement with those of the REM simulation (data not shown).

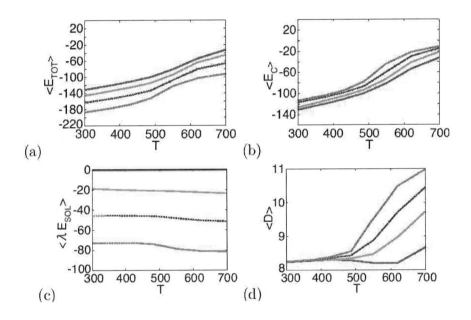

Figure 4.2. The average total energy (a), average conformational energy (b), average of $\lambda \times E_{SOL}$ (c), and average end-to-end distance (d) with all the λ values as functions of temperature. The lines colored in red, green, blue, and purple are for λ_1, λ_2, λ_3, and λ_4, respectively. They are in order from above to below in (a) and (c) and from below to above in (b) and (d).

We found that the results of the two-dimensional ST simulation are in complete agreement with those of the two-dimensional REM simulation for the average quantities. The only difference between the two simulations is the number of replicas. In the present simulation, while the REM simulation used 32 replicas, the ST simulation used only one replica. Hence, we can save much computer power with ST.

A second example of our multidimensional generalized-ensemble simulations is a pressure ST (PST) simulation in the isobaric-isothermal ensemble [30]. This simulation performs a random walk in one-dimensional pressure space. The system that we simulated is ubiquitin in explicit water. This system has been studied by high pressure NMR experiments and known to undergo high-pressure denaturations [40, 41]. Ubiquitin has 76 amino acids and it was placed in a cubic box of 6232 water molecules. Temperature was fixed to be 300 K throughout the simulations, and we prepared 100 values of pressure ranging from 1 bar to 10000 bar. Temperature and pressure were controlled by Hoover-Langevin method [42] and particle mesh Ewald method [43, 44] were employed for electrostatic interactions. The time step was 2.0 fsec. The force field CHARMM22 [45] with CMAP [46, 47] and TIP3P water model [45, 48] were used, and the program package NAMD version 2.7b3 [49] was modified to incorporate the PST algorithm.

We first performed 100 independent conventional isobaric-isothermal simulations of 4 nsec with $T = 300$ K (i.e., $M_0 = 1$) and 100 values of pressure (i.e., $M_1 = 100$). Using the obtained histogram $N_{m_0,m_1}(E, \mathcal{V})$ of potential energy and volume distribution, we obtained the ST parameters f_{m_0,m_1} by solving the WHAM equations in Eqs. (4.2.14) and (4.2.15). We then performed the PST production of 500 nsec and repeated it 10 times with different seeds for random numbers (so, the total simulation time for the production run is 5.0 μsec).

In Fig. 3 we show the time series of pressure and potential energy during the PST production run. In the Figure we see a random walk in pressure between 1 bar and 10000 bar. A random walk in potential energy is also observed and it is anti-correlated with that of pressure, as it should be.

We calculated the fluctuations $\sqrt{<d^2> - <d>^2}$ of the distance d between pairs of C^α atoms. The results are shown in Fig. 4. We see that large fluctuations are observed between residues around 7-10 and around 20-40, which are in accord with the experimental results [40, 41].

The fluctuating distance corresponds to that between the turn region of the β-hairpin and the end of the α-helix as depicted in Fig. 5. While at low pressure

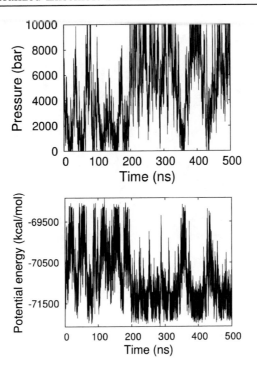

Figure 4.3. Time series of pressure (left) and potential energy (right) during the PST production run.

this distance is small, at high pressure it is larger and water comes into the created open region.

4.4. Conclusion

In this Chapter we presented the multidimensional extensions of two well-known generalized-ensemble algorithms, namely, REM and ST, which can greatly enhance conformational sampling of biomolecular systems. We generalized the original potential energy function E_0 by adding any physical quantities V_ℓ of interest as a new energy term with a coupling constant $\lambda^{(\ell)}(\ell = 1, \cdots, L)$. The simulations in multidimensional REM and multidimensional ST algorithms realize a random walk in temperature and $\lambda^{(\ell)}(\ell = 1, \cdots, L)$ spaces.

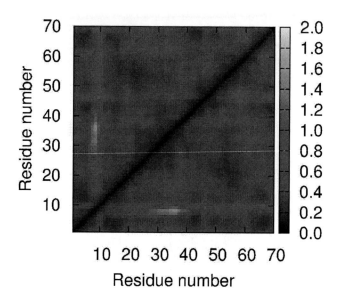

Figure 4.4. Fluctuations of distance between pairs of C^α atoms that was calculated from the PST production run.

While the multidimensional REM simulation can be easily performed because no weight factor determination is necessary, the required number of replicas can be quite large and computationally demanding. We thus prefer to use the multidimensional ST, where only a single replica is simulated, instead of REM. However, it is very difficult to obtain optimal weight factors for the multidimensional ST. Here, we presented a powerful method to determine the weight factors. Namely, we first perform a short multidimensional REM simulation and use the multiple-histogram reweighting techniques to determine the weight factors for the multidimensional ST simulations.

The multidimensional generalized-ensemble algorithms that were presented in the present Chapter will be very useful for Monte Carlo and molecular dynamics simulations in materials and biological sciences.

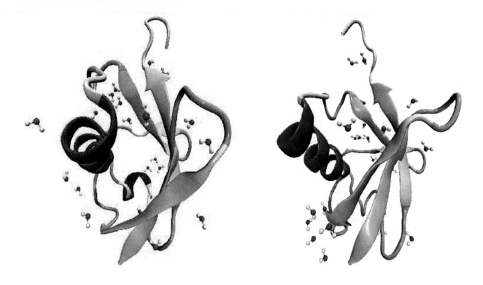

Figure 4.5. Snapshots of ubiquitin during the PST production run at low pressure (left) and at high pressure (right).

Acknowledgment

Some of the results were obtained by the computations on the super computers at the Institute for Molecular Science, Okazaki, and the Institute for Solid State Physics, University of Tokyo, Japan. This work was supported, in part, by Grants-in-Aid for Scientific Research on Innovative Areas ("Fluctuations and Biological Functions"), and for the Next-Generation Super Computing Project, Nanoscience Program from the Ministry of Education, Culture, Sports, Science and Technology (MEXT), Japan.

References

[1] Kirkpatrick, S., Gelatt, C.D. Jr., and Vecchi, M.P. (1983) Optimization by simulated annealing. *Science* **220**, 671–680.

[2] Nilges, M., Clore, G.M., and Gronenborn, A.M. (1988) Determination of three-dimensional structures of proteins from interproton distance data by hybrid distance geometry-dynamical simulated annealing calculations. *FEBS Lett.* **229**, 317–324.

[3] Brünger, A.T. (1988) Crystallographic refinement by simulated annealing. Application to a 2.8 Åresolution structure of aspartate aminotransferase. *J. Mol. Biol.* **203**, 803–816.

[4] Wilson, S.R., Cui, W., Moskowitz, J.W., and Schmidt, K.E. (1988) Conformational analysis of flexible molecules - location of the global minimum energy conformation by the simulated annealing method. *Tetrahedron Lett.* **29**, 4373–4376.

[5] Kawai, H., Kikuchi, T., and Okamoto, Y. (1989) A prediction of tertiary structures of peptide by the Monte Carlo simulated annealing method. *Protein Eng.* **3**, 85–94.

[6] Wilson, C. and Doniach, S. (1989) A computer model to dynamically simulate protein folding: studies with crambin. *Proteins* **6**, 193–209.

[7] Kawai, H., Okamoto, Y., Fukugita, M., Nakazawa, T., and Kikuchi, T. (1991) Prediction of α-helix folding of isolated C-peptide of ribonuclease A by Monte Calro simulated annealing. *Chem. Lett.* **1991**, 213–216.

[8] Okamoto, Y., Fukugita, M., Nakazawa, T., and Kawai, H. (1991) α-Helix folding by Monte Carlo simulated annealing in isolated C-peptide of ribonuclease A. *Protein Eng.* **4**, 639–647.

[9] Hansmann, U.H.E. and Okamoto, Y. (1999) New Monte Carlo algorithms for protein folding. *Curr. Opin. Struct. Biol.* **9**, 177-183.

[10] Mitsutake, A., Sugita, Y., and Okamoto, Y. (2001) Generalized-ensemble algorithms for molecular simulations of biopolymers. *Biopolymers* **60**, 96–123.

[11] Sugita, Y. and Okamoto, Y. (2002) Free-energy calculations in protein folding by generalized-ensemble algorithms. In: Schlick, T. and Gan, H.H. Eds. *Lecture Notes in Computational Science and Engineering*, (Springer-Verlag, Berlin) pp. 304–332; e-print: cond-mat/0102296.

[12] Okamoto, Y. (2004) Generalized-ensemble algorithms: enhanced sampling techniques for Monte Carlo and molecular dynamics simulations. *J. Mol. Graphics Mod.* **22**, 425–439; e-print: cond-mat/0308360.

[13] Itoh, S.G., Okumura, H., and Okamoto, Y. (2007) Generalized-ensemble algorithms for molecular dynamics simulations. *Mol. Sim.* **33**, 47–56.

[14] Sugita, Y., Mitsutake, A., and Okamoto, Y. (2008) Generalized-ensemble algorithms for protein folding simulations. In: Janke, W. Ed. *Lecture Notes in Physics. Rugged Free Energy Landscapes: Common Computational Approaches in Spin Glasses,Structural Glasses and Biological Macromolecules*, (Springer-Verlag, Berlin) pp. 369–407; e-print: arXiv:0707.3382v1[cond-mat.stat-mech].

[15] Okamoto, Y. (2009) Generalized-ensemble algorithms for studying protein folding. In: Kuwajima, K., Goto, Y., Hirata, F., Kataoka, M., and Terazima, M. Eds. *Water and Biomolecules*, (Springer-Verlag, Berlin) pp. 61–95.

[16] Ferrenberg, A.M. and Swendsen, R.H. (1988) New Monte Carlo technique for studying phase transitions. *Phys. Rev. Lett.* **61**, 2635–2638; *ibid.* **63**, 1658 (1989).

[17] Ferrenberg, A.M. and Swendsen, R.H. (1989) Optimized Monte Carlo data analysis. *Phys. Rev. Lett.* **63**, 1195–1198.

[18] Kumar, S., Bouzida, D., Swendsen, R.H., Kollman, P.A.,and Rosenberg, J.M. (1992) The weighted histogram analysis method for free-energy calculations on biomolecules. 1. The method. *J. Comput. Chem.* **13**, 1011–1021.

[19] Hukushima, K. and Nemoto, K. (1996) Exchange Monte Carlo method and application to spin glass simulations. *J. Phys. Soc. Jpn.* **65**, 1604–1608.

[20] Marinari, E., Parisi, G., and Ruiz-Lorenzo, J.J. (1997) Numerical simulations of spin glass systems. In: Young, A.P. Ed. *Spin Glasses and Random Fields*, (World Scientific, Singapore) pp. 59–98.

[21] Sugita, Y. and Okamoto, Y. (1999) Replica-exchange molecular dynamics method for protein folding. *Chem. Phys. Lett.* **314**, 141–151.

[22] Lyubartsev, A.P., Martinovski, A.A., Shevkunov, S.V., and Vorontsov-Velyaminov, P.N. (1992) New approach to Monte Carlo calculation of the free energy - method of expanded ensemble. *J. Chem. Phys.* **96**, 1776–1783.

[23] Marinari, E. and Parisi, G. (1992) Simulated tempering - a new Monte Carlo scheme. *Europhys. Lett.* **19**, 451–458.

[24] Sugita, Y., Kitao, A., and Okamoto, Y. (2000) Multidimensional replica-exchange method for free-energy calculations. *J. Chem. Phys.* **113**, 6042–6051.

[25] Fukunishi, F., Watanabe, O., and Takada, S. (2002) On the Hamiltonian replica exchange method for efficient sampling of biomolecular systems: Application to protein structure prediction. *J. Chem. Phys.* **116**, 9058–9067.

[26] Mitsutake, A. and Okamoto, Y. (2009) From multidimensional replica-exchange method to multidimensional multicanonical algorithm and simulated tempering. *Phys. Rev. E* **79**, 047701.

[27] Mitsutake, A. and Okamoto, Y. (2009) Multidimensional generalized-ensemble algorithms for complex systems. *J. Chem. Phys.* **130**, 214105.

[28] Mitsutake, A. (2009) Simulated-tempering replica-exchange method for the multidimensional version. *J. Chem. Phys.* **131**, 094105.

[29] Okumura, H. and Okamoto, Y. (2006) Multibaric-multithermal ensemble molecular dynamics simulations. *J. Comput. Chem.* **27**, 379–395.

[30] Mori, Y. and Okamoto, Y. (2010) Generalized-ensemble algorithms for the isobaric-isothermal ensemble. *J. Phys. Soc. Jpn.* **79**, 074003.

[31] Torrie, G.M. and Valleau, J.P. (1977) Nonphysical sampling distributions in Monte Carlo free-energy estimation: Umbrella sampling. *J. Comput. Phys.* **23**, 187–199.

[32] Mitsutake, A. and Okamoto, Y. (2000) Replica-exchange simulated tempering method for simulations of frustrated systems. *Chem. Phys. Lett.* **332**, 131–138.

[33] Myers, J.K., Pace, C.N., and Scholtz, J.M. (1997) A direct comparison of helix propensity in proteins and peptides. *Proc. Natl. Acad. Sci. U.S.A.* **94**, 2833–2837.

[34] Mitsutake, A., Sugita, Y., and Okamoto, Y. (2003) Replica-exchange multicanonical and multicanonical replica-exchange Monte Carlo simulations of peptides. II. Application to a more complex system. *J. Chem. Phys.* **118**, 6676–6688.

[35] Momany, F.A., McGuire, R.F., Burgess, A.W., and Scheraga, H.A. (1975) Energy parameters in polypeptides. VII. Geometric parameters, partial atomic charges, nonbonded interactions, hydrogen bond interactions, and intrinsic torsional potentials for the naturally occurring amino acids. *J. Phys. Chem.* **79**, 2361–2381.

[36] Némethy, G., Pottle, M.S., and Scheraga, H.A. (1983) Energy parameters in polypeptides. 9. Updating of geometrical parameters, nonbonded interactions, and hydrogen bond interactions for the naturally occurring amino acids. *J. Phys. Chem.* **87**, 1883–1887.

[37] Sippl, M.J., Némethy, G., and Scheraga, H.A. (1984) Intermolecular potentials from crystal data. 6. Determination of empirical potentials for O-H...O=C hydrogen bonds from packing configurations. *J. Phys. Chem.* **88**, 6231–6233.

[38] Ooi, T., Oobatake, M., Némethy, G., and Scheraga, H.A. (1987) Accessible surface areas as a measure of the thermodynamic parameters of hydration of peptides. *Proc. Natl. Acad. Sci. U.S.A.* **84**, 3086–3090.

[39] Masuya, M., unpublished; see http://biocomputing.cc/nsol/.

[40] Kitahara, R. and Akasaka, K. (2003) Close identity of a pressure-stabilized intermediate with a kinetic intermediate in protein folding. *Proc. Natl. Acad. Sci. U.S.A.* **100**, 3167–3172.

[41] Kitahara, R., Yokoyama, S., and Akasaka, K. (2005) NMR snapshots of a fluctuating protein structure: ubiquitin at 30 bar - 3 kbar. *J. Mol. Biol.* **347**, 277–285.

[42] Quigley, D. and Probert, M.I.J. (2004) Landevin dynamics in constant pressure extended systems. *J. Chem. Phys.* **120**, 11432–11441.

[43] Darden, T., York, D., and Pedersen, L. (1993) Particle mesh Ewald - an Nlog(N) method for Ewald sums in large systems. *J. Chem. Phys.* **98**, 10089–10092.

[44] Essmann, U., Perera, L., Berkowitz, M.L., Darden, T., Lee, H., and Pedersen, L.G. (1995) A smooth particle mesh Ewald method. *J. Chem. Phys.* **103**, 8577–8593.

[45] MacKerell, A.D. Jr., Bashford, D., Bellott, M., Dunbrack, R.L. Jr., Evanseck, J.D., Field, M.J., Fischer, S., Gao, J., Guo, H., Ha, S., Joseph-McCarthy, D., Kuchnir, L., Kuczera, K., Lau, F.T.K., Mattos, C., Michnick, S., Ngo, T., Nguyen, D.T., Prodhom, B., Reiher, W.E. III, Roux, B., Schlenkrich, M., Smith, J.C., Stote, R., Straub, J., Watanabe, M., Wiórkiewicz-Kuczera, J., Yin, D., and Karplus, M. (1998) All-atom empirical potential for molecular modeling and dynamics studies of proteins. *J. Phys. Chem. B* **102**, 3586–3616.

[46] MacKerell, A.D. Jr., Feig, M., and Brooks, C.L. III (2004) Improved treatment of the protein backbone in empirical force fields. *J. Am. Chem. Soc.* **126**, 698–699.

[47] MacKerell, A.D. Jr., Feig, M., and Brooks, C.L. III (2004) Extending the treatment of backbone energetics in protein force fields: Limitations of

gas-phase quantum mechanics in reproducing protein conformational distributions in molecular dynamics simulations. *J. Comput. Chem.* **25**, 1400–1415.

[48] Jorgensen, W.L., Chandrasekhar, J., Madura, J.D., Impey, R.W., Klein, M.L. (1983) Comparison of simple potential functions for simulating liquid water. *J. Chem. Phys.* **79**, 926–935.

[49] Phillips, J.C., Braun, R., Wang, W., Gumbart, J., Tajkhorshid, E., Villa, E., Chipot, C., Skeel, R.D., Kale, L., and Schulten, K. (2005) Scalable molecular dynamics with NAMD. *J. Comput. Chem.* **26**, 1781–1802.

In: Molecular Dynamics of Nanobiostructures   ISBN: 978-1-61324-320-6
Editor: K. Kholmurodov   © 2012 Nova Science Publishers, Inc.

Chapter 5

LATERAL HETEROGENEITY AS AN INTRINSIC PROPERTY OF HYDRATED LIPID BILAYERS: A MOLECULAR DYNAMICS STUDY

Darya V. Pyrkova[a], Natalya K. Tarasova[a,b],
Nikolay A. Krylov[a,c], Dmitry E. Nolde[a]
and Roman G. Efremov[a,*]

[a]M.M. Shemyakin & Yu.A. Ovchinnikov Institute of
Bioorganic Chemistry
Russian Academy of Sciences, Moscow, Russia;
[b]Department of Bioengineering, Biological Faculty
M.V. Lomonosov Moscow State University, Moscow, Russia
[c]Joint Supercomputer Center
Russian Academy of Sciences, Moscow, Russia

Abstract

Studies of lateral heterogeneity in cell membranes are important since they help to understand the physical origin of lipid domains and rafts.

---

[*]E-mail address: efremov@nmr.ru

The simplest membrane mimics are hydrated bilayers composed of saturated and unsaturated lipids. While their atomic structural details resist easy experimental characterization, important insight can be gained *via* computer modeling. We present the results of all-atom molecular dynamics simulations for a series of fluid one- and two-component hydrated lipid bilayers composed of phosphatidylcholines with saturated (dipalmitoylphosphatidylcholine, DPPC) and mono-unsaturated (dioleoylphosphatidylcholine, DOPC) acyl chains slightly differing in length (16 and 18 carbon atoms, respectively). As a results, it was shown that the bilayers' properties are tuned in a wide range by the chemical nature and relative content of lipids. The impact that the micro-heterogeneity may have on formation of lateral domains in response to external signals is discussed. Understanding of such effects creates a basis for rational design of artificial membranes with predefined properties.

Keywords: Lateral heterogeneity, Lipid bilayers, Molecular dynamics study

5.1.   Introduction

Apart from the barrier function separating contents of cells or cellular compartments from the exterior, lipid bilayer of biological membranes plays a critical role in numerous biochemical processes in the living organisms. Instead to be a passive "sea" with polar surfaces and a hydrophobic core, where different proteins and other molecules can accommodate their functions, multicomponent lipid bilayers of cell membranes represent themselves a highly active, dynamic, and self-organizing medium [1]. The simplest systems mimicking cellular membrane are lipid bilayers composed of different types of lipids. Understanding on molecular level the main trends in their structural and dynamic organization under different conditions may shed light on behavior of real membranes. The lateral arrangement of lipids of different types in multicomponent bilayers was proved important for physico-chemical properties of the bilayer itself, as well as for structure and functioning of membrane proteins [2]. In particular, the distribution of lipid components in the plane of membrane is not homogeneous - ideal mixing occurs only in certain cases [3,4]. Such lateral heterogeneities in lipid bilayers differ in their spatial dimensions. Thus, in the current discussion the major accent is done on lipid rafts [1] - the large-scale ($\sim$1000 Å) clusters enriched in particular lipids, sterols, and proteins. Much smaller (100-500 Å) domains have been registered in experiments with model mixed bilayers [5]. As a rule, lipid mixing-demixing phenomena are most apparent when coupled to a phase transition for the gel-fluid coexistence. On the contrary, far from the critical points, correlation lengths of lipids in liquid phases usually are of the order of a molecular size [6], thus leading to formation of the so-called "microdomains" [7] including only a few lipids or proteins. It is important to note that heterogeneities of this size exist in the liquid disordered (l_d) phase, which corresponds to the native conditions of the living organism. The most appropriate system allowing characterization of "microdomains" in the fluid state is a bilayer composed of two phospholipids - under such conditions, many of them demonstrate non-ideal mixing [6,8-10]. Such "microdomains" lie below the level of resolution by modern experimental techniques. Nevertheless, it was shown that the lateral lipid distribution deviates from that specific for ideal binary solution, while the most interesting atomic-scale details of such systems resist easy experimental characterization. Further progress in studies of mixed lipid bilayers in the fluid state can be achieved via computer simulation techniques. One of the most informative among them is molecular dynamics (MD). It is therefore interesting to assess in detail structural and dynamic properties of the two-component mixtures of lipids. Despite importance of the problem, such systems have been investigated in much less extent as compared with lipid-cholesterol mixtures. In most of these studies, lipid molecules with the same acyl chains but different head groups were considered [11-15]. A nonlinear change of properties at different concentrations of components in the systems was detected [11-12]. By contrast, only few MD simulations were reported for bilayers containing phospholipids with different chains [8,16-18]. Moreover, mixtures of saturated lipids were considered in most cases. These studies have established that some binary lipid systems [5,8,16-18] form non-ideal mixtures exhibiting positive deviations from ideality. It was shown that the gel/liquid crystalline phase transition temperature T_m depends on the concentration of lipids. To our knowledge, at the present moment there are only few simulation works on mixed all-atom membranes containing lipids with similar head groups, but differing in the saturation state of acyl chains (see, e.g [19]). Such interactions could play important role in biological membranes. It is well known that the ratio of saturated and unsaturated acyl chains differs not only between mammalians and plants [20-22] or from one species to another, but can even vary greatly for the organs of one species [23,24]. In addition, a number of experimental studies on phosphatidylcholine mixtures containing saturated and mono-unsaturated chains in the liquid phase have also been reported (see [25,26] for recent reviews). While in the early studies of binary systems it was supposed that PCs behave similarly regardless of the exact nature of their acyl chains [4], the more recent results demonstrated that even relatively modest differences in the fatty acid structure of phospholipids can produce lateral heterogeneity in the fluid phase, in the absence of other added compounds [9,10,27,28]. Taking the above said into account, in this work we present the results of all-atom molecular dynamics simulations carried out for a series of one- and two-component lipid bilayers composed of dioleoylphosphatidylcholine (DOPC) and dipalmitoylphosphatidylcholine (DPPC). The mixed systems were considered in l_d phase at eight different molar ratios of lipids.

5.2.   Materials and Methods

5.2.1.   Construction of Bilayer Systems, MD-Simulation Protocol

MD simulations were performed on symmetric (according to composition of monolayers) fully hydrated DOPC/DPPC systems containing 288 lipid molecules (144 per leaflet) arranged in a bilayer structure. The force field for DPPC was taken from Berger et. al. [29], and that for DOPC was elaborated based on the parameters available for POPC (http://moose.bio.ucalgary.ca/ tieleman/Download.html) [30]. The Ryckaert-Bellemans torsion potential was used for the lipid hydrocarbon chains [31]. Nonbonded interactions were described by the parameters from Berger et. al. [29] Partial atomic charges were obtained from Chiu et al. [32] In the beginning of the simulations, the lipids were distributed randomly in the mixed systems. The bilayer models were then placed into rectangular boxes and solvated with SPC [33] water molecules. The united-atom representation was used for the methyl/metylene groups in the acyl chains. Prior to MD simulations, all systems were subjected to energy minimization (1000 conjugate gradients steps). Each system was subsequently heated to 325 K during 60 ps. Then, 15 ns collection MD runs were carried out for all considered systems, except DOPC and DOPC50. For the latter ones, 30-ns MD trajectories were obtained. All simulations were performed with the GROMACS (v 3.3.1) simulation suite [34]. Simulations were carried out with a time step of 2 fs, with imposed 3D periodic boundary conditions, in the NPT ensemble with an isotropic pressure of 1 bar and the temperature 325 K, which is above the gel/liquid-crystal phase transition temperature ($T_m \approx 315$ K for DPPC, $T_m \approx 250$ K for DOPC) [28]. All components of the systems (water, lipids) were coupled separately to a temperature bath using Berendsen thermostat [35] with a coupling constant of 0.1 ps. van der Waals interactions were truncated using the twin-range 12/20 Å spherical cutoff function. Electrostatic interactions were treated with the Particle Mesh Ewald (PME) algorithm [36] (1.2 Å Fourier spacing).

5.2.2. Analysis of MD Trajectories

The computed 15-ns MD trajectories were analyzed with the help of original software and modified utilities supplied with the GROMACS package. To compare the simulation results with the experimental data, several important macroscopic averages were estimated for the near-equilibrium parts (last 5 ns) of the trajectories. The area per lipid molecule (A_L) was calculated from the cross-sectional area of the simulation box divided by the number of lipids per leaflet (144). Bilayer thickness ($\langle D_{P-P} \rangle$) was taken as a distance between mean positions of phosphorus atoms of the opposite bilayer leaflets.

5.2.3. Analysis of MHP Distribution on the Bilayer Surface

Solvent-accessible surface area (ASA) of the membranes was calculated for MD snapshots of the bilayer model using the PLATINUM program [37]. All calculations related to mapping and visualization of MHP properties were done as described in [37,38]. Depending on the MHP-value, each surface point was attributed either to "hydrophilic" (MHP$\leq$-0.1), "hydrophobic" (MHP$\geq$0.1) or "neutral" (-0.1 $<$ MHP $<$ 0.1) one.

5.2.4. Cluster Analysis

For geometrical cluster analysis we estimated pairwise distances (r) between atoms P8 in polar heads of lipids. Molecules with r $<$ r_c = 6 Å were combined in a cluster. A lipid was considered to be a part of the cluster if its distance with any other molecule in the cluster was smaller than r_c. Time - dependent behavior of lipids in clusters was analyzed as follows. If a lipid belongs to the same cluster during a considerably large period of time (more than 0.7 ns), than this lipid and the corresponding cluster were taken into account.

5.3. Results and Discussion

5.3.1. Overall Properties of One- and Two-Component Lipid Bilayers

5.3.1.1. Structural Parameters

Average area per lipid (A_L) and bilayer thickness ($\langle D_{P-P} \rangle$) are among the most important structural parameters. For pure DOPC and DPPC systems, the average values of A_L and $\langle D_{P-P} \rangle$ are only slightly lower than the experimental ones [39,40,41-43]. This makes us confident that the force field and MD-protocols employed provide a reasonable representation of the DOPC and DPPC bilayers. The calculated values of A_L and $\langle D_{P-P} \rangle$ demonstrate a sharp transition upon addition of a small amount of saturated DPPC into the DOPC bilayer (Fig. 1). At higher concentrations of DPPC, there is a linear dependence of A_L on the lipid composition. For all mixed bilayers, the values of A_L are smaller than it was expected in case of ideal lipid mixing in the systems. On the other hand, bilayer thickness in the two component systems is almost the same as in the pure DPPC. Overall, it is seen that the macroscopic averages characterizing geometry of the binary systems are much close to saturated than to unsaturated bilayer.

5.3.1.2. Hydrophobic/Hydrophilic Properties of Water-Bilayer Interfaces

Hydrophobic organization of the water-bilayer interfaces was assessed in terms of MHP values calculated in each surface point (see Methods). Such mapping reveals that the water-lipid interface is not purely hydrophilic as one could expect based on a "traditional" model of membrane organization, where the hydrophobic core is shielded from water by polar lipid heads. Instead, the bilayer surface contains considerable amount of hydrophobic zones accessible to water. Such a "mosaic" hydrophobic/hydrophilic character of surfaces in lipid membranes has been revealed in our MD simulations of other water-lipid systems [44,45]. Detailed inspection of the DOPC/DPPC bilayers shows that up to 90% of their hydrophobic surface is created by the lipid acyl chains "snorkeling" near the membrane-water interface. Therefore, the hydrophobic "spots" correspond to zones between the lipid heads. The higher is the total hydrophobic area, the looser is lipid packing because the area per PC head undergoes only minor changes and may be considered constant. Obviously, the hydrophobic area in a lipid bilayer should correlate with the A_L values. Thus, the maximal and mini-

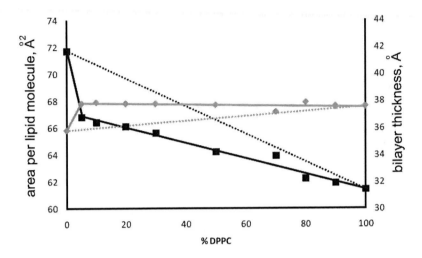

Figure 5.1. Average area per lipid molecule (black squares and lines) and bilayer thickness (grey rhombuses and lines) vs DPPC content in DOPC, DPPC, and DOPC/DPPC bilayers calculated based on MD data (solid line) and expected for linear dependence on lipid composition (dotted line).

mal hydrophobic areas correspond to pure DOPC and DPPC, respectively. The same is true for their A_L values (Fig. 1).

5.3.2.   Clustering in Lipid Bilayers as Manifestation of Packing and Hydrophobic Effects

Detailed inspection of lipid packing reveals that geometrical clusters are formed in all bilayer systems under study. It is noteworthy that most of the clusters have lifetimes of approximately 1 - 2 ns. Clusters that persist for more than 2 ns are very rare. In addition, spatial and size distributions of clusters before and after MD simulations are completely different. Thus, in the beginning of the MD runs, no clusters were detected in the one-component bilayers, and no clusters containing three or more lipids were found in the mixed bilayers. This shows that possible dependence of lateral lipid distributions on initial state is removed to a large extent at the final stages of MD, near the equilibrium. Prior to analysis of the differences in lipid properties in the clusters and in the rest of the bilayer, typical size and composition of clusters were estimated. The total number of clustered lipids varies from 50 to 75 molecules, which is nearly a quarter of all molecules in the bilayer (288 lipids). The clusters however are not large - their most preferable size is three lipids per cluster. At the same moment, only a negligible part of clusters is composed of two lipids. This indicates that formation of clusters is not an occasion. Also, there is a number of larger clusters comprising four or five lipid molecules. The reason of the small cluster size is their instability in time, which, in turn, is the result of rather weak intermolecular interactions in those systems - in the liquid disordered state, the corresponding free energy for DOPC/DPPC mixture is +70 cal/mol [25]. In all binary systems, the number of clusters containing both types of lipid molecules (hereinafter referred to as "mixed" clusters) is nearly the same and represents about 45% of the total amount of clusters. With further change of lipid concentrations, only the composition of "mixed" clusters varies, but not their relative content among all clusters. It could be explained by slight preference of DOPC and DPPC to be surrounded by alike lipid molecules, which however does not completely exclude unlike neighbors. Lipids form one-component clusters only if their concentration in the bilayer exceeds 40-50%. At smaller concentrations, all minor lipid component susceptible to tight packing, is incorporated into the "mixed" clusters. To assess in more details the differences in behavior of lipids upon clustering, all lipid molecules in each frame of the simulation trajectory were attributed to one of the three groups: "clustered", "interchanging", and "non-clustered". For example, lipids in clusters were divided into two groups. Those that were changing their partners in clusters fall into group "interchanging". And those that had the same neighbors in consequent MD-frames fall in the group "clustered" (see Methods).

5.3.2.1. Hydrophobic/Hydrophilic Properties of Lipid Bilayer Surface: Role of Clustering

To get deeper insight into the surface properties of hydrated lipid bilayers, in-plane distributions of hydrophobic/hydrophilic patterns for "clustered", "interchanging", and "non-clustered" lipid groups were calculated (Fig. 2). This was done using the MHP approach (see Methods). The results were compared with those for the entire bilayer. In all the systems, quite similar regularities were observed. As seen in Figure 2, clustered lipids have slightly less hydrophobic surface than the bilayer in average. The opposite effect is observed for lipids

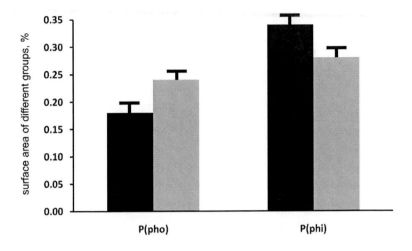

Figure 5.2. Average fractions of hydrophobic P(pho) and hydrophilic P(phi) surface area occupied by "clustered" (black) and "non-clustered" (gray) lipids in the hydrated DOPC bilayer.

outside clusters. No significant deviations from the average bilayer properties are found for lipids in the group "interchanging". The compact arrangement of heads prevents "surfacing" of dynamically flexible lipid chains responsible for formation of hydrophobic regions on the surface. That is why surface of the geometrical clusters is more hydrophilic. On the contrary, heads of the lipids outside clusters are positioned far from each other. Therefore, their chains meet less resistance for "snorkeling".

5.3.3.   Intrinsic Heterogeneity in Model Lipid Bilayers: Microscopic Origin and Putative Biological Role

Based on the presented results of MD analysis of a series of one- and two-component bilayers composed of PCs with saturated (DPPC) and mono-unsaturated (DOPC) acyl chains slightly differing in length (16 and 18 carbon atoms, respectively), the following conclusions can be drawn. (1) Lateral arrangement of lipids in all studied systems is not random. Instead, it reveals occurrence of small (most often, 3 lipids) clusters on the surface. Clustering seems to be a feature inherent in the lipid bilayers near the equilibrium state. (2) In mixed bilayers, the clusters are formed both, by alike and unlike lipids. One-component clusters appear only at high concentrations (above 40-50%) of this type of lipid. (3) The observed microheterogeneity picture is highly dynamic - lipid molecules enter and leave clusters with the characteristic lifetimes about 1 ns. (4) Clustering determines to a large extent thestructural, dynamic, and hydrophobic properties of bilayers, like area per lipid molecule, bilayer thickness, order parameters of acyl chains (data not shown), parameters of the mosaic hydrophobicity patterns on the lipid-water interface, and so forth. (5) Addition of even small (5-10%) amount of DPPC to the DOPC bilayer strongly affects the physico-chemical characteristics of the entire system, while the opposite is not true. By other words, in the two-component system, one (DPPC) may be considered as an "order-preferring" agent, which efficiently modulates behavior of the second one (DOPC).

5.3.4. Atomic-Scale Origin of Lipid Clustering

Analysis of MD data for one- and two-component zwitterionic lipid bilayers clearly shows that the structural/dynamic/hydrophobic organization of a water-lipid interface is determined by a thin balance of differently directed factors. On the one hand, avoidance of lipids to pack together permits efficient solvation of their polar heads with water. Such a process is energetically favorable. At the same time, this leads to creation of the "free space" on the surface, which is immediately filled by "snorkeling" fragments of acyl chains, thus resulting in appearance of hydrophobic zones exposed to water and, hence, to increasing the total free energy of the system. Next, slightly energetically unfavorable clustering of DOPC and/or DPPC lipids is compensated by formation of polar patches on the water-accessible surface of the membrane because this increases conformational entropy of the interfacial water. In equilibrium, the balance between different energy terms and therefore stability of the bilayer is maintained (among other factors) due to fast dynamic processes - creation and destroying of small clusters accompanied with rearrangements of solvation shells around lipid heads, and so forth. It is worth noting that the small size of clusters is important - large domains are too inert to provide fast relaxation and adequate response on external perturbations, whichtend to shift the equilibrium in the system.

5.3.5.   Putative Biological Impact from Simulations of Model Systems

It seems reasonable to suggest that such a complicated character defining physico-chemical equilibrium properties of even simplest lipid bilayers may have prominent biological impact. In living cells, there is strong evidence for dynamic raft-based membrane heterogeneity at the nanoscale, which can be functionally coalesced to more stable membrane-ordered assemblies [2]. In most of recent studies, an emphasis is made on such large-scale membrane perturbations and formation of rafts, which play significant role in membrane protein localization and modulating of their functioning. These objects and processes have been detected in numerous experiments. Apart from lipids, such systems contain "order-preferring components" (usually, cholesterol) inducing clustering, phase separation, and so on. Furthermore, global structural rearrangements in cell membranes are often caused by the action of external agents, like proteins and/or particular low-molecular-weight compounds (anesthetics, steroids, etc.). On the other hand, lipid-protein, etc. interactions alone cannot describe lipid rafts, because these do not account for the preferentially connecting lipid-lipid interactions that have been demonstrated in model membranes even without any "ordering" compounds [2]. In other words, lipid bilayer of cell membranes is not just an inert matrixaccommodating functional modules like proteins. Instead, due to the physically selective behavior of lipids in a chemically specific manner, even simple one- and two-component hydrated bilayer systems in a liquid state already reveal organized micro-heterogeneity. Upon extrapolation from model system to living cells, one can suppose that such effects make biomembranes ready for fast and adequate formation of lateral domains in response to external signals. Understanding of the molecular mechanisms determining the lateral heterogeneous organization of the water-lipid systems makes it possible rational computational design of new membrane structures with particular "mosaic" patterns defined by structural, dynamic, and hydrophobic properties of the systems. Hopefully, such bilayers will possess specific functional characteristics required for modern biomedical applications, e.g. facilitating binding of molecules (including drugs and drug targets) of a given size, specific 3D hydrophobic/electrostatic properties, and so on.

Acknowledgments

This work was supported by the Russian Foundation for Basic Research and by the RAS Programmes (MCB and "Basic fundamental research of nanotechnologies and nanomaterials"). Access to computational facilities of the Joint Supercomputer Center (Moscow) and Moscow State University is gratefully acknowledged.

References

[1] Lingwood, D., Kaiser, H.-J., Levental, I., Simons, K., "Lipid rafts as functional heterogeneity in cell membranes". *Biochem. Soc. Trans.*, **37**, (2009), 955-960.

[2] Lingwood, D., Simons, K., "Lipid rafts as a membrae-organizing principle". *Science*, **327**, (2010), 46-50.

[3] Freire, E., Snyder, B. "Estimation of the lateral distribution of molecules in two-component lipid bilayers". *Biochemistry*, **19**, (1980), 88-94.

[4] Curatolo, W., Sears, B., Neuringer, L.J., "A calorimetry and deuterium NMR study of mixed model membranes of 1-palmitoyl-2-oleylphosphatidylcholine and saturated phosphatidylcholines". *Biochim. Biophys. Acta.*, **817**, (1985), 261-270.

[5] Sugar, P., Michonova-Alexova, E., Chong, P.L.-G., "Geometrical properties of gel and fluid clusters in DMPC/DSPC bilayers: Monte Carlo simulation approach using a two-state model". *Biophys. J.*, **81**, (2001), 2425-2441.

[6] Spaar, A., Salditt, T., "Short range order of hydrocarbon chains in fluid phospholipid bilayers studied by X-Ray diffraction from highly oriented membranes". *Biophys. J.*, **85**, (2003), 1576-1584.

[7] Freire, E., Snyder, B. "Estimation of the lateral distribution of molecules in two-component lipid bilayers". *Biochemistry*, **19**, (1980), 88-94.

[8] de Joannis, J., Jiang, Y., Yin, F., Kindt, J.T., "Equilibrium distributions of dipalmitoyl phosphatidylcholine and dilauroyl phosphatidylcholine in

a mixed lipid bilayer: atomistic Semigrand canonical ensemble simulations". *J. Phys. Chem. B.*, **110**, (2006), 25875-25882.

[9] Dewa, T., Vigmond, S.J., Regen, S.L., "Lateral heterogeneity in fluid bilayers composed of saturated and unsaturated phospholipids". *J. Am. Chem. Soc.*, **118**, (1996), 3435-3440.

[10] Risselada, H.J., Marrink, S.J., "The molecular face of lipid rafts in model membranes". *Proc. Natl. Acad*, **105**, (2008), 17367-17372.

[11] Leekumjorn, S., Sum, A.K., "Molecular simulation study of structural and dynamic properties of mixed DPPC/DPPE bilayers". *Biophys. J.*, **90**, (2006), 3951-3965.

[12] de Vries, H., Mark, A.E., Marrink, S.J., "The binary mixing behavior of phospholipids in a bilayer: A molecular dynamics study". *J. Phys. Chem. B.*, **108**, (2004), 2454-2463.

[13] Pandit, S.A., Bostick, D., Berkowitz, M.L., "Mixed bilayer containing dipalmitoylphosphatidylcholine and dipalmitoylphosphatidylserine: lipid complexation, ion binding, and electrostatics". *Biophys. J.*, **85**, (2003), 3120-3131.

[14] López Cascales, J.J., Otero, T.F., Smith, B.D., González, C., Marquez, M., "Model of an asymmetric DPPC/DPPS membrane: effect of asymmetry on the lipid properties. A molecular dynamics simulation study". *J. Phys. Chem. B.*, **110**, (2006), 2358-2363.

[15] Murzyn, K., Rg, T., Pasenkiewicz-Gierula, M., "Phosphatidylethanol-amine-phosphatidylglycerol bilayer as a model of the inner bacterial membrane". *Biophys. J.*, **88**, (2005), 1091-1103.

[16] Faller, R. Marrink, S.L., "Simulation of domain formation in DLPC-DSPC mixed bilayers". *Langmuir*, **20**, (2004), 7686-7693.

[17] Bennun, S.V., Longoand, M., Faller, R., "Phase and mixing behavior in two-component lipid bilayers: A molecular dynamics study in DLPC/DSPC mixtures". *J. Phys. Chem. B.*, **111**, (2007), 9504-9512.

[18] Wang, H., de Joannis, J., Jiang, Y., Gaulding, J.C., Albrecht, B., Yin, F., Khanna, K., Kindt, J.T., "Bilayer edge and curvature effects on partitioning

of lipids by tail lenght: atomistic simulations". *Biophys. J.*, **95**, (2008), 2647-2657.

[19] Róg, T., Murzyn, K., Gurbiel, R., Takaoka, Y., Kusumi, A., Pasenkiewicz-Gierula, M., "Effects of phospholipid ansaturation on the bilayer nonpolar region: a molecular simulation study". *J. Lipid Res.*, **45**, (2004), 326-336.

[20] Perona, J.S., Vögler, O., Sánchez-Domínguez, J.M., Montero, E., Escribá, P.V., Ruiz-Gutierrez, V., Ruiz-Gutierrez, J., "Consumption of virgin olive oil influences membrane lipid composition and regulates intracellular signaling in elderly adults with type 2 diabetes mellitus". *A Biol. Sci. Med. Sci.*, **62**, (2007), 256-263.

[21] Yoo, S.K., Brown, I., Gaugler, R., "Liquid media development for hetero-rhabditis bacteriophora: lipid source and concentration". *Biotechnol.*, **54**, (2000), 759-763.

[22] Fernandez, M.L., Lin, E.C., McNamara, D.J., "Regulation of guinea pig plasma low density lipoprotein kinetics by dietary fat saturation". *J. Lipid. Res.*, **33**, (1992), 97-109.

[23] Fryer, L.G.D., Orfali, K.A., Holness, M.J., Saggerson, E.D., Sugden, M.C., "The long-term regulation of skeletal muscle pyruvate dehydrogenase kinase by dietary lipid is dependent on fatty acid composition". *Eur. J. Biochem.*, **229**, (1995), 741-748.

[24] Vazquez-Memije, M.E., Maria, J., Cardenas-Mendez, M.J., Tolosa, A., Hafidi, M.E., "Respiratory chain complexes and membrane fatty acids composition in rat testis mitochondria throughout development and ageing". *Exp. Gerontol.*, **40**, (2005), 482-490.

[25] Almeida, P.F.F., "Thermodynamics of lipid interactions in complex bilayers". *Biochim. Biophys. Acta.*, **1788**, (2009), 72-85.

[26] Veatch, S.L., Keller, S.L., "Seeing spots: complex phase behavior in simple membranes". *Biochim. Biophs. Acta*, **1746**, (2005), 172-185.

[27] Scherfeld, D., Kahya, N., Schwille, P., "Lipid dinamics and domain formation in model membranes composed of trenary mixtures of unsaturated and saturated phosphatidylcholines and cholesterol". *J. Chem. Phys.*, **85**, (2003), 3758-3768.

[28] Schmidt, M.L., Ziani, L., Boureau, M., Davis, J.H., "Phase equilibria in DOPC/DPPC: conversion from gel to subgel in two component mixtures". *J. Chem. Phys.*, **131**, (2009), 175103.

[29] Berger, O., Edholm, O., Jahning, F., "Molecular dynamics simulations of a fluid bilayer of dipalmitoylphosphatidylcholine at full hydration, constant pressure, and constant temperature". *Biophys. J.*, **72**, (1997), 2002-2013.

[30] Tieleman, D.P., Sansom, S.M., Berendsen, H.J.C., "Alamethicin helices in a bilayer and in solution: molecular dynamics study". *Biophys J.*, **76**, (1999), 40-49.

[31] Ryckaert, J.P., Bellemans, A., "Molecular dynamics of liquid n-butane near its boiling point". *Chem. Phys. Lett.*, **30**, (1975), 123-125.

[32] Chiu, S.W., Clark, M., Balaji, V., Subramaniam, S., Scott, H.L., Jakobsson, E., "Incorporation of surface tension into molecular dynamics simulation of an interface: a fluid phase lipid bilayer membrane". *Biophys. J.*, **69**, (1995), 1230-1245.

[33] Berendsen, H.J.C., Postma, J.P.M., van Gunsteren, W.F., Hermans, J., "Interaction Models for Water in Relation to Protein Hydration". *in. Pullman, B., Ed.; Intermolecular Forces;*, (1981).

[34] Lindahl, E., Hess, B., van der Spoel, D., "A package for molecular simulation and trajectory analysis". *J. Mol. Mod.*, **7**, (2001), 306-317.

[35] Berendsen, H.J.C., Postma, J.P.M., van Gunsteren, DiNola, A., Haak, J.R., "Molecular dynamics with coupling to an external bath". *J. Chem. Phys.*, **81**, (1984), 3684-3690.

[36] Essmann, U., Perera, L., Berkowitz, M.L., Darden, T., Lee, H., Pedersen, L.G., "A smooth particle mesh ewald potential". *J. Chem. Phys.*, **103**, (1995), 8577-8592.

[37] Pyrkov, T.V., Chugunov, A.O., Krylov, N.A., Nolde, D.E., Efremov, R.G., "PLATINUM: a web tool for analysis of hydrophobic/hydrophilic organization of biomolecular complexes". *Bioinformatics*, **25**, (2009), 1201-1202.

[38] Efremov, R.G., Vergoten, G., "The hydrophobic nature of membrane-spanning a-helices as revealed by Monte Carlo simulations and molecular hydrophobicity potential analysis". *J. Physical Chemistry*, **99**, (1995), 10658-10666.

[39] Tristram-Nagle, S., Petrache, H.I., Nagle, J.F., "Structure and interactions of fully hydrated dioleoylphosphatidylcholine bilayers". *Biophys. J.*, **75**, (1998), 917-925.

[40] Petrache, H.I., Dodd, S.W., Brown, M.F., "Area per lipid and acyl length distributions in fluid phosphatidylcholines determined by 2H NMR spectroscopy". *Biophys. J.*, **79**, (2000), 3172-3192.

[41] Nagle, J.F., Tristram-Nagle, S., "Structure of lipid bilayers". *Biochim. Biophys. Acta.*, **1469**, (2000), 159-195.

[42] Kucerka, N., Nagle, J.F., Sachs, J.N., Feller, S.E., Pencer, J., Jackson, A., Katsaras, J., "Lipid bilayer structure determined by the simultaneous analysis of neutron and X-ray scattering data". *Biophys. J.*, **95**, (2008), 2356-2367.

[43] Kucerka, N., Tristram-Nagle, S., Nagle, J.F., "Look at structure of fully hydrated fluid phase DPPC bilayers". *Biophys. J.*, **90**, (2006), 83-85.

[44] Polyansky, A.A., Volynsky, P.E., Arseniev, A.S., Efremov, R.G., "Adaptation of a membrane-active peptide to heterogeneous environment. II. The role of mosaic nature of the membrane surface". *J. Phys. Chem. B.*, **113**, (2009), 1120-1126.

[45] Polyansky, A.A., Volynsky, P.E., Nolde, D.E, Arseniev, A.S., Efremov, R.G., "Role of lipid charge in organization of water/lipid bilayer interface: Insights via computer simulations". *J. Phys. Chem. B.*, **109**, (2005), 15052-15059.

In: Molecular Dynamics of Nanobiostructures   ISBN: 978-1-61324-320-6
Editor: K. Kholmurodov   © 2012 Nova Science Publishers, Inc.

Chapter 6

LARGE MODELING THE CHOLESTERIC PHASE OF THE DNA

Albina A. Sevenyuk[1], Veniamin N. Blinov[1],
*Voislav L. Golo[1] and Konstantin V. Shaitan[2, *]*
Moscow State University,
[1]Department of Mechanics and Mathematics,
[2]Department of Biology
Moscow, Russia

Abstract

We study the cholesteric liquid crystallin phases of the DNA, and employ a computer simulation that relies on the electrostatical picture of the interaction between molecules of the DNA in solution. To that end we use a qualitative model that describes a molecule of the DNA as a one-dimensional lattice framed by charges due the phosphate groups, and dipoles of base pairs subject to the helicoidal symmetry. Our results are in agreement, by orders of magnitude, with the experimental data.

Keywords: Computer simulation, DNA, Cholesteric phase

---
[*]E-mail address: kireeva.al@gmail.com

6.1. Introduction

The liquid crystalline phases of solutions of the DNA, [1], promise new technological applications, [2], [3], [4], [5], [6]. There are several types of liquid crystals, [7]. Nematic liquid crystals are characterized by the orientational order determined by the axes of molecules, considered as cylindrical rods. Cholesteric liquid crystals have the helicoidal structure; the molecules of a cholesteric crystal having a common orientation in two dimensional layers, the orientation changing from a layer to the adjacent one by a rotation about an axis through an angle Φ, or cholesteric pitch. The liquid crystalline phases of the DNA have one more degree of freedom due to the helical structure of molecules of the DNA. The relative position of their molecules may differ by the rotation about the axis of cholesteric through the pitch angle Φ, and also by a rotation about the axis of the duplex of the DNA through an angle ϕ determined by positions of adjacent base pairs of a molecule, see Fig.6.1.

It is important, that the formation of liquid crystalline phases of the DNA requires the segments of the DNA being short enough, approximately 150 bp, or less. This is due to the fact that a molecule of the DNA has the persistence length approximately 500 Å, depending on the ionic strength of solution, [14], so that the molecule may be visualized as a straight rod. It is worth noting that the theoretical treatment of *nematic*, and to a large extent cholesteric, liquid crystals, is based on the picture of an ensemble of hard rods, [8]. But cholesteric liquid crystals require a more sophisticated treatment, [7]. Their theory is still in the state of development, see paper [9].

The theoretical study of the cholesteric phases of the DNA relies on the pair interactions of constituent molecules. By now it is generally accepted that it is mainly of electrostatic origin. Thus, Kornyshev and Leikin, [6], put forward the theory that relies on the helical distribution of charges in a molecule of the DNA. They consider a molecule of the DNA as a charged rod or cylinder, the charge being distributed continuously over the surface of the rod, subject to the helicoidal symmetry of the duplex. The investigation of the pair interaction obtained in this way, indicates that the orientational order of the molecules should be the cholesteric one.

So far, only the pair interaction has been considered, and the thermodynamics of ensembles of molecules of the DNA still requires investigation. The reason lies in that the analytical treatment, which has been employed, runs across serious difficulties in considering ensembles. Thus, the problem needs a com-

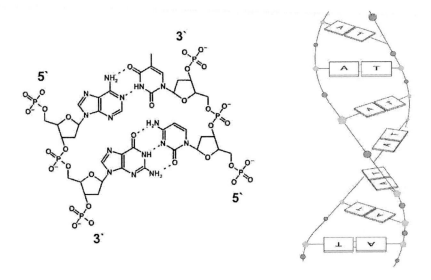

Figure 6.1. Steric picture of a fragment of the DNA, left. Helical turn of the molecule, right.

puter analysis, but the latter requires a formulation using discrete terms. To that end we have constructed a lattice model that accommodates the helical structure and the distribution of charges of a molecule of the DNA. Our work is based mainly on the picture of electrostatic interaction of paper [6]. Aiming at numerical estimates for the existence of cholesteric phases in solutions of the DNA, we may perform the numerical simulations of liquid crystallin phases of the DNA, and study ensemble of molecules by means of discrete electrostatic model. Earlier such approach was employed in the study of pair interaction [12], [13].

6.2. A Lattice Model of a Charged Molecule of the DNA

The DNA is highly charged molecule owing to the phosphate groups and the dipole moments of the base pairs [11]. Therefore, we consider a molecule of the DNA as 1-dimensional lattice of charges and dipoles. The size of an elementary cell equals 3.4 Å, which mimics the spatial conformation of the B-form of the DNA. The dipoles are perpendicular to the lattice axis, and their conformation has the helical symmetry corresponding to the helical structure of DNA, the charges lying half-way between adjacent dipoles. Since the structure of a molecule of the DNA is not regular due to various factors, the sequence of base pairs, for example, our assumption of the regular positions of the dipole and the charges, is only an approximation.

Figure 6.2. Discrete model of the DNA molecule: the dipoles (arrows) are perpendicular to the lattice axis; the helix symmetry with the pitch $\pi/5$. Charges (circles) half-way between adjacent dipoles

In estimating the values of the charges and the dipoles one has to take into account that they are to a large extent screened by ions of the solution, or counterions, and thus effectively neutralized. Bloomfield and coworkers, [14], found that the charges of the DNA must be neutralized by more than 90% in order that cholesteric LC-phases could be formed. To that end they usually employ spermine+3, spermidine+4 solutions to obtain liquid crystalline phases. In our calculations the value of charge equals 5% of the charge of the phosphate group. The results of our calculations suggest the phosphate charges and the dipoles of base pairs should be neutralized differently.

The following dimensional units are useful for the calculations we perform:

- mass $M = 10^{-22}\, gr$;

- length $L = 3 \times 10^{-8}\, cm$;

- time $T = 10^{-13}\, sec$;

Employing the above units we obtain the value of the electron charge, $1\,Debye$, the dipole moment and the energy $1\,erg$

$$1\,e = 1, \quad 1, \quad Debye = 0.06, \quad 1\,erg = 10^{11}$$

According to [11], electrical dipole moments of the bases varies within $1 \div 10 Debye$, what is corresponding to $0.06 \div 0.6$ in the above system of units. Similarly, in the above units the dipole moment is considered to be 0.6, and the neutralized chargee 0.05, correspondingly.

The pair interaction energy is the sum of elementary interaction energies:

$$U = \sum (u_{dd} + u_{cd} + u_{dc} + u_{cc})$$

where u_{dd} is the energy of the interaction between the dipole of the first molecule and the dipole of the second, u_{cd} is the energy of the interaction between the charge of the first molecule and the dipole of the second, u_{dc} is the energy of the interaction between the dipole of the first molecule and the charge of the second, u_{cc} is the energy of the interaction between the charge of the first molecule and the charge of the second. We have

$$u_{dd} = \frac{1}{\rho^3} \vec{p} \cdot \vec{p}' - \frac{3}{\rho^5} (\vec{p} \cdot (\vec{r} - \vec{r}')) (\vec{p}' \cdot (\vec{r} - \vec{r}'))$$

$$u_{dc} = \frac{Q'}{\rho^3} \vec{p} \cdot (\vec{r}' - \vec{r})$$

$$u_{cd} = \frac{Q}{\rho^3} \vec{p}' \cdot (\vec{r} - \vec{r}')$$

$$u_{cc} = \frac{Q \cdot Q'}{\rho}$$

$$\rho = |\vec{r} - \vec{r}'|$$

where $\vec{r}$, $\vec{r}'$ are vectors from the origin to the corresponding lattice sites, Q, Q' charges of the corresponding phosphate groups.

The important point about the electrostatic interaction is the screening factor. It is not clear whether one may employ the Debye-Hückel theory to study solutions of the DNA. Since our model is rough, we have taken into account the screening by changing the values of charges used in calculations. The results obtained in this way indicate that the neutralization of the phosphate charges is important for the formation of the liquid crystalline phases. We have used short segments of 32 base pairs because the calculations employing segments of 152 base pairs, as in experimental papers [4], [2], are too time consuming. But a few test calculations with long chains allow us to suggest that our results are in a good qualitative agreement with the data of papers [4], [2].

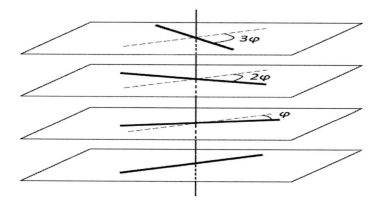

Figure 6.3. Conformation of the molecules of the ensemble. The molecules lie in parallel plains, the distances between the plains are equal and fixed.

6.3.   The Numerical Simulation

We study an ensemble of molecules of the DNA in the Gibbs thermostat, and aim at finding the equilibrium conformation. Our main instrument is the Monte Carlo calculations using the Metropolis algorithm. We have used parallel computations with the GPU card and CUDA architecture.

The problem is still too hard to solve in its general setting owing to the lack of computer power. To get round this obstacle we consider a specific conformation that has sufficient structure so as to see whether a cholesteric phase of the DNA may be treated within the framework of the electrostatic model defined above. Thus, we suppose that the molecules lie in parallel plains, their centers being fixed on a straight line, perpendicular to the plains. The distance Δ between adjacent plains equals 14 Å, in accord with the experimental data of papers [4], [5], [2], by the order of magnitude. One may consider Δ as a parameter describing the concentration of molecules of the DNA in solution. The cholesteric conformation of the molecules is illustrated in Fig. 6.3.

The key point in constructing the Markov chain for the Metropolice algorithm is to accommodate the degrees of freedom of the framed lattice describing a molecule of the DNA. We employed a random generator that rotate the lattice (1) through a small random angle about its axis and (2) through another small random angle about the common axis of the ensemble, see Fig.6.3. Generally, we have employed the Markov chains $N = 10^5$ steps long, and for the control

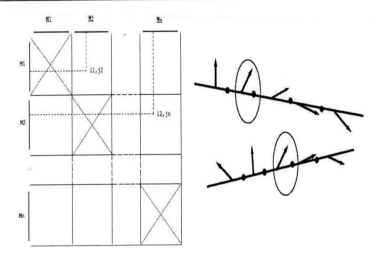

Figure 6.4. Calculation of the pair potential. The lattice of a molecule is divided into cells. The cells are indicated by $M_1, M_2, \ldots$. A cell is formed by the charge corresponding to a site of the lattice and the dipole on the right-hand side of the charge. Each thread block of the GPU is responsible for computing the interaction of a cell of one molecule with a cell of another. The shared memory is not used.

$N = 10^6$ steps long. The dispersion was about $D = 10^{-3}$, so that the precision, $\sqrt{D}/N$, is equal to 10^{-5}. The details of parallel computations with the GPU card and CUDA architecture are illustrated in the Fig. 6.4.

Our results are shown in Figs.6.5 and 6.6. We see that the chiral angle describing the rotation of molecule about the common axis increases linearly, Fig.6.5. Thus, we have obtained a cholesteric structure. Deviations from the chiral pitch, Fig.6.6, are by an order of magnitude smaller than the chiral angle itself. Therefore, we may conclude that the cholesteric structure may suffer conformational fluctuations, but remains unbroken but heat fluctuations.

6.4. Summary

1. The electrostatic theory of cholesteric phase of the DNA, [6], considered within the framework of the discrete model of framed lattices, provides good qualitative agreement with the experimental data, [4], [2]. The calculations are

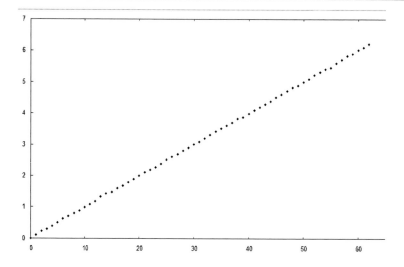

Figure 6.5. Chiral angle, radians. Dependence on the molecule number. Temperature 300 K

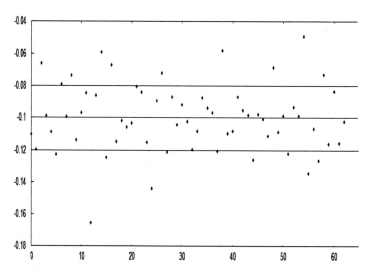

Figure 6.6. Deflection angle of the neighboring molecules. Dependence on the molecule number. Temperature 300 K

in agreement with the neutralization of the charges of the DNA necessary for the formation of the cholesteric phase, first indicated in [14].

2. The screening effect needs further clarification. The familiar Debye-Hükel approach seems to be too simplistic for solutions of the DNA.

Acknowledgment

We used the computing clusters GPU C1060-Tesla, the Department of Biology, and the Computer Center, the Lomonosov Moscow State university. The work has been partially supported by the grants the Scientific Schools NSh-3224.2010.1 and the Russian Foundation for Fundamental research (RFFI) 09—02—00551a.

References

[1] C. Robinson, "Liquid-crystalline structures in polypeptide solutions". *Tetrahedron*, **13**, (1961), 219-234.

[2] Yu.M. Yevdokimov, "Liquid crystalline forms of DNA and their biological role". *Liquid Crystals (Russian)* **3**, (2003), 10.

[3] Yu.M. Yevdokimov, V.I.Salyanov, E.V. Shtykova, K.A. Dembo, V.V. Volkov ,P.V. Spirin, A.S. Slusheva, V.S. Prassolov, "A Transition in DNA Molecule's Spatial Ordering Due to Nano-Scale Structural Changes". *The Open Nanoscience Journal*, **2**, (2008), 17-28.

[4] F. Livolant, A. Leforestier, "Condensed Phases of DNA: Structures and Phase Transitions". *Prog. Polym. Sci.*, **21**, (1996), 1115-1164.

[5] E. Raspaud, D. Durand, F. Livolant, "Interhelical Spacing in Liquid Crystalline Spermine and Spermidine-DNA Precipitates". *Biophys. J.*, **88**, (2005), 392-403.

[6] A.A. Kornyshev, S. Leikin, "Theory of interaction between helical molecules". *J. Chem. Phys.*, **107**, (1997), 3656-3674.

[7] S. Chandrasekhar, "Liquid Crystals". *Ch. 4, Cambridge University press*, (1977).

[8] L. Onsager, "The effects of shape on the interaction of colloidal particles". *Ann. Acad. Sci.*, **51**, (1949), 627-659.

[9] T.C. Lubensky, "A spin model for cholesteric liquid crystals". *J. Phys. Chemistry of Solids*, **34**, (1973), 365-370.

[10] C.T. Zhang, "Harmonic and subharmonic resonances of microwave absorption in DNA". *Phys. Rev.*, **A40**, (1989), 2148-2153.

[11] J. Sponer, J. Leszcynski, P. Hobza, "Electronic properties, hydrogen bonding, stacking, and cation binding of DNA and RNA bases". *Biopolymers*, **61**, (2002), 3-31.

[12] V.L. Golo, E.I. Kats, Yu.S. Volkov, "Symmetries of Electrostatic Interaction between DNA Molecules". *Pisma ZhETF*, **86**, (2007), 311-316.

[13] V.L. Golo, E.I. Kats, S.A. Kuznetsova, Yu.S. Volkov, "Symmetry of electrostatic interaction between pyrophosphate DNA molecules". *Eur. Phys. J. E*, **31**, (2010), 59-67.

[14] V.A. Bloomfield, "DNA condensation". *Curr. Opin. Struct. Biol.*, **6**, (1996), 334-341.

In: Molecular Dynamics of Nanobiostructures   ISBN: 978-1-61324-320-6
Editor: K. Kholmurodov   © 2012 Nova Science Publishers, Inc.

Chapter 7

Ab Initio MOLECULAR DYNAMICS STUDY OF DISORDERED MATERIALS

Satoshi Ohmura and Fuyuki Shimojo[*]
Department of Physics, Kumamoto University,
Kumamoto 860-8555, Japan

Abstract

The atomic dynamics of liquid B_2O_3 and the energy-transfer mechanism in light harvesting dendrimers are studied by *ab initio* molecular-dynamics simulations. In liquid B_2O_3, it is found that the diffusivity of boron becomes about two times larger than that of oxygen under pressure above 20 GPa while the former is 10-20 % smaller than the latter at lower pressures. We reveal the microscopic origin of this anomalous pressure dependence of diffusivity. We take into account the nonadiabatic effects in molecular-dynamics simulations to describe the photo-exitation state of dendrimers. Our simulation reveals the key role of thermal molecular motion that significantly accelerates the energy transport based on the Dexter mechanism.

Keywords: *Ab initio*, Molecular dynamics, Disordered materials

---
[*]E-mail address: 095d9003@st.kumamoto-u.ac.jp

7.1. Introduction

In recent years, the material research has increasingly focused on understanding the dynamic and electronic properties of disordered materials. Due to the deviation from the perfect order in the atomic structure, they exhibit scientifically and technologically important phenomena. One of the interesting disordered materials is liquid B_2O_3, which is the typical covalent liquid. The transport properties of covalent liquids under pressure are very important, because they are closely related with igneous processes in the Earth. For a number of covalent liquids, such as SiO_2, GeO_2, and silicates, abnormal behavior of the viscosity has been observed, i.e., the viscosity significantly drops with pressure. [1, 2] Recently, *in situ* viscosity measurements of liquid B_2O_3 under high pressure have been reported. [3] It was shown that the viscosity deceases with increasing pressure, as in other covalent liquids. Since the viscosity is considered to be related to the atomic diffusion, it is of particular interest to investigate the pressure dependence of the microscopic-diffusion mechanism in liquid B_2O_3. It is, however, unclear how the rearrangement process of the covalent bonds is affected by compression.

Synthetic supermolecules such as light-harvesting dendrimers are attracting great attention because harvesting energy from sunlight is of paramount importance for the solution of the global energy problem. [4] In these molecules, the electronic excitation energy due to photoexcitation of antennas located on the periphery of the molecules is rapidly transported to the photochemical reaction centers at the cores of the molecules, which in turn performs useful work such as photosynthesis and molecular actuation. [5] A number of experimental [6–8] and theoretical [9–13] works have addressed rapid energy transport mechanisms in light-harvesting dendrimers. Such energy transfer is conventionally attributed to either dipole-dipole interactions (Förster mechanism) [14, 15] or the overlapping of donor and acceptor electronic wave functions (Dexter mechanism). However, atomistic mechanisms of rapid electron transport in these dendrimers remain elusive.

In this paper, we report on detailed investigations of the pressure dependence of atomic diffusion in liquid B_2O_3 [16, 17] and microscopic mechanisms of energy transport in a light-harvesting dendrimer by *ab initio* molecular-dynamics (MD) simulations, in which interatomic forces are calculated quantum mechanically in the framework of density-functional theory (DFT). The purposes of our study are to clarify the microscopic mechanism of atomic diffusion in covalent liquids under pressure and to elucidate the effect of atomic motion on rapid energy transport after photoexcitation of light-harvesting dendrimers.

7.2. Method of Calculation

The electronic states are calculated using the projector-augmented-wave (PAW) method [18,19] within the framework of DFT. The generalized gradient approximation (GGA) [20] is used for the exchange-correlation energy. The plane-wave cutoff energies are 30 and 200 Ry for the electronic pseudo-wave functions and the pseudo-charge density, respectively. The Γ point is used for Brillouin zone sampling.

For the simulations of liquid B_2O_3, a system of 120 (48B+72O) atoms in a cubic supercell is used under periodic boundary conditions. The equations of motion for atoms are solved via an explicit reversible integrator [21] with a time step of $\Delta t = 1.2$ fs. To determine the density of the liquid state under pressure, a constant-pressure MD simulation [22] is performed for 2.4 ps at each given pressure. Using the time-averaged density, the static and diffusion properties are investigated by MD simulations in the canonical ensemble. [23] Seven thermodynamic states used in this study cover a density range from 1.69 to 3.99 g/cm^3 and a pressure range from 1.4 to 97.0 GPa, as listed in Table 7.1. The temperature of 2500 K is chosen so as to be sufficiently high to maintain the liquid state even at the high pressures investigated. The procedure for obtaining the liquid state from the crystalline configuration has been described in a previous paper. [26] The quantities of interest are obtained by averaging over 21.6 ps long enough to achieve good statistics after the initial equilibration, which takes at least 2.4 ps.

For the simulation of the dendrimer, the system consists of a zinc-porphyrin core and a benzyl ether-type antenna. In the antenna, there are three aromatic rings connected by ether oxygen atoms, out of which one aromatic ring is directly connected to the zinc-porphyrin core. The periodic boundary condition is employed with a supercell of dimension $18 \times 18 \times 24$ Å which is large enough to avoid the interaction between periodic images of the molecule. A time step of $\Delta t = 0.48$ fs is used. The nonadiabatic MD simulations are carried out by incorporating electronic transitions through the fewest-switches surface-hopping (FSSH) method [27] along with the Kohn-Sham (KS) representation of time-dependent (TD) DFT. The nuclei are treated classically in the adiabatic rep-

Table 7.1. Densities ρ (g/cm^3) used in MD simulations of liquid B$_2$O$_3$ in the canonical ensemble at a temperature of 2500 K. The relative volumes V/V_0, where V_0 is the volume at $\rho = 1.50$ g/cm^3 ($P = 0.0$ GPa), and the time-averaged pressures [24, 25] P (GPa) are also listed

ρ (g/cm^3)	V/V_0	P (GPa)
1.69	0.89	1.4
1.92	0.78	3.2
2.36	0.64	9.2
2.63	0.57	15.5
3.08	0.49	28.4
3.42	0.44	46.2
3.99	0.38	97.0

resentation, i.e., the atomic forces are calculated from the (excited) electronic eigenstates for the current nuclear positions. Switching probability from the current adiabatic state to another is computed from the density-matrix elements obtained by solving the TDKS equations, [28] and nonadiabatic transitions between adiabatic states occur stochastically. [27]

7.3. Dynamic Properties of Liquid B$_2$O$_3$ under Pressure

7.3.1. Coordination Number Distribution and Diffusion Coefficients

In order to discuss the pressure dependence of the local structure in detail, we show the pressure dependence of the coordination-number distribution $f_{\alpha-\beta}^{(n)}$ for $\alpha - \beta =$ B-O and O-B in Figs.1(a) and 1(b), respectively. $f_{\alpha-\beta}^{(n)}$ is the ratio of the number of α-type atoms that are coordinated to n β-type atoms to the total number of α-type atoms. To obtain $f_{\alpha-\beta}^{(n)}$, we count the number of β-type atoms inside a sphere with radius R centered at each α-type atom, where $R = 1.9$ Å is a cutoff distance determined from the first-minimum position of the pair distribution function $g_{B-O}(r)$ at 1.4 GPa. Both $f_{B-O}^{(3)}$ and $f_{O-B}^{(2)}$ have high values of about 0.98, i.e., 98% of B and O atoms have threefold and twofold coordination, respectively, under pressure up to about 3 GPa. With increasing pressure, for $P > 3$ GPa, $f_{B-O}^{(3)}$ and $f_{O-B}^{(2)}$ decrease, and instead $f_{B-O}^{(4)}$ and $f_{O-B}^{(3)}$ increase. While $f_{B-O}^{(3)}$ and $f_{B-O}^{(4)}$ are exchanged for each other at about 25 GPa, $f_{O-B}^{(2)}$ and $f_{O-B}^{(3)}$ are interchanged at a higher pressure of about 50 GPa because of the composition ratio of B and O atoms. At about 100 GPa, the number of fourfold-coordinated B atoms approaches about 90%, and about 5% of B atoms have fivefold coordination.

7.3.2. Pressure Dependence of the Diffusion Coefficients

Figure 7.1 (c) shows the diffusion coefficients d_α for $\alpha = $ B and O atoms as a function of pressure. Clearly, liquid B_2O_3 has a diffusion maximum around 10 GPa. In the pressure range of $0 < P < 10$ GPa, both d_B and d_O show a similar increase with increasing pressure, which is consistent with the observed pressure dependence of the viscosity of the undercooled liquid. [29] The diffusion enhancement is related to a weakening of the covalent-like interaction between atoms accompanying the increase in the coordination number, as in other covalent liquids. [30, 31] In fact, we found that in liquid B_2O_3 under pressures above 3 GPa, long-range atomic diffusion occurs by the usual concerted reactions, while the nonbridging O atoms are always involved in diffusion processes at lower pressures. [26] In each concerted reaction, two BO_4 groups are generated as an intermediate by forming two new B-O bonds between adjoining BO_3 units, whereas only one BO_4 group is required to produce one nonbridging O atom. [17] The concerted reactions occur more frequently with increasing pressure, which should be originated from covalent-bond weakening due to compression, and should enhance the atomic diffusion. We also observed that the reactions with the nonbridging O atoms decrease in number with pressure and almost disappear for $P > 10$ GPa.

The concerted reactions as well as those with the nonbridging O atoms involve equal numbers of B and O atoms; therefore, d_B and d_O would show a similar pressure dependence up to about 10 GPa, as seen in Fig. 7.1 (c). The decrease in the diffusivity above a certain pressure is not surprising but quite natural. It is, however, unusual that the diffusivity of O atoms is reduced more quickly than that of B atoms with compression above 10 GPa, which indicates another diffusion mechanism at such high pressures. As a result, d_B becomes

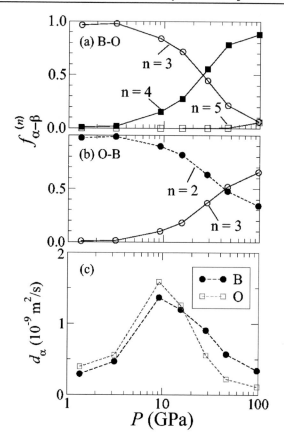

Figure 7.1. Pressure dependence of the coordination-number distribution $f^{(n)}_{\alpha-\beta}$ for $\alpha-\beta$ = (a) B-O and (b) O-B. (c) Pressure dependence of the diffusion coefficients d_α for α = B (solid circles) and O (open squares) atoms.

about two times larger than d_O when the pressure exceeds 20 GPa. Note that such pressure dependence has not been observed in the first-principles study of liquid SiO_2. [30]

7.3.3. Diffusion Mechanism under High Pressure

It is worth noting that the number of fourfold-coordinated B atoms becomes larger than that of threefold-coordinated B atoms to consider the anomalous dif-

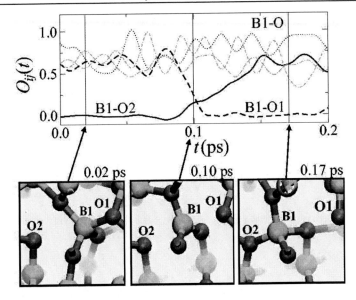

Figure 7.2. (Top panel) The time evolution of bond-overlap populations $O_{ij}(t)$ for $i = $ B1, $j \in$ O in the diffusion process observed at 46.2 GPa. The thick solid and thick dashed lines show $O_{ij}(t)$ associated with the B and O atoms of interest. The thin lines show $O_{ij}(t)$ between the B atom of interest (labeled as 'B1') and their neighboring O atoms except O1 and O2. (Bottom panel) Atomic configurations at $t = 0.02, 0.10$, and 0.17 ps. The large and small spheres show B and O atoms, respectively.

fusive properties of liquid B_2O_3 under pressure. We focus on diffusion processes associated with fourfold-coordinated B atoms. A typical example is shown in Fig. 7.2, where the time evolution of the bond-overlap populations, $O_{ij}(t)$, associated with the B and O atoms of interest is displayed with snapshots of atomic configurations. $O_{ij}(t)$ yields a semiquantitative estimate of the strength of the covalent-like bonding between atoms. [32, 33] In the beginning of this process, the B atom labeled 'B1' has fourfold coordination. The atomic configuration at 0.02 ps (and also at 0.17 ps) shows that fourfold-coordinated B atoms have a tetrahedral arrangement, which demands sp^3 hybridization around them. As displayed in the top panel of Fig. 7.2, $O_{B1-O1}(t)$ begins to decrease at about 0.08 ps and almost vanishes for $t > 0.11$ ps. This time change means that the covalent bond between B1 and O1 is broken within 0.03 ps. On the other hand, $O_{B1-O2}(t)$ increases gradually between 0.08 and 0.15 ps, indicating that a covalent bond is formed between B1 and O2, taking 0.07 ps to form. As shown in the atomic configuration at 0.10 ps, the B1 atom has threefold coordination with a planar arrangement before the formation of the new B1-O2 bond is complete. In this way, one of the B-O bonds is broken quickly, and a new B-O bond is formed gradually when B atoms move between the fourfold-coordinated sites.

In the diffusion process shown in Fig. 7.2, twofold-coordinated O atoms, toward which fourfold-coordinated B atoms move, are necessary as B1 moves toward O2. Note that O2 has twofold-coordination before B1 bonds to it. As shown in Fig. 7.1(b), twofold-coordinated O atoms exist even at high pressures. This fact indicates that this diffusion process occurs rather frequently, and that B atoms can diffuse fairly easily. On the other hand, for the concerted reaction involving the migration of O and B atoms, threefold-coordinated B and twofold-coordinated O atoms are necessary. However, the number of threefold-coordinated B atoms decreases rapidly with increasing pressure for $P > 20$ GPa, which means that the concerted reaction is suppressed at such high pressures. This is why d_O decreases rapidly with increasing pressure.

7.4. Electron Transport in Dendrimers

7.4.1. Spatial Distribution of Wave Functions

Figure 7.3 shows the spatial distribution of one-electron wave functions of the lowest unoccupied molecular orbital (LUMO) and LUMO+4 in the ground state, where the atomic positions are relaxed so as to minimize the total energy. It is seen from Fig. 7.3 that LUMO spread only within the core. The highest occupied molecular orbital (HOMO), as well as LUMO+1, LUMO+2, HOMO-1, HOMO-2 distribute also only around the core. These spatial distributions of the electronic states near HOMO and LUMO are consistent with the fact that electrons and holes photoexcited in the peripheries eventually move to the core. In contrast to these core states, LUMO+4 spread mainly within the peripheries. The wave functions of HOMO-3, HOMO-4, and LUMO+3 are also distributed mainly within the peripheries.

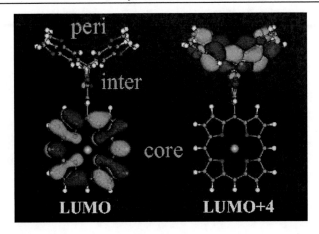

Figure 7.3. Spatial distribution of electronic wave functions in the ground state, for LUMO and LUMO+4, where red and green colors represent the isosurfaces of the wave functions with the values of 0.013 and −0.013 a.u., respectively.

7.4.2. Time Evolution of Electronic Eigenenergies in the Ground State

Figure 7.4 shows the time evolution of electronic eigenenergies ε_i during the MD simulation. Even at a finite temperature of 300 K, the two core states, LUMO and LUMO+1, are not mixed with the other states. Figure 7.4 exhibits multiple crossings of eigenenergies in this energy range. When an eigenenergy is well separated from the others, its wave function has a large amplitude only within the core or one of the peripheries. On the other hand, the wave function spreads over both core and a periphery, when the eigenenergies crossing or approach each other. This suggests that electrons photoexcited in the peripheries are transferred to the core through such extended state, i.e., by the Dexter mechanism. A similar situation is observed for the occupied states (HOMO, HOMO-1, ...), which suggests that hole transport also occurs with the same mechanism.

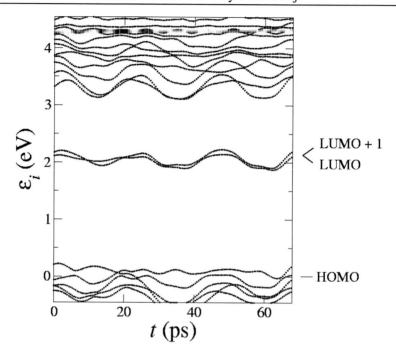

Figure 7.4. Time evolution of electronic eigenenergies during adiabatic MD simulation for the ground state.

7.4.3. Time Evolution of Electronic Eigenenergies in Excitation State

In order to confirm that a photoexcited electron indeed transfers from the peripheries to the core based on the Dexter mechanism, we perform nonadiabatic MD simulations by using the TDKS-FSSH methods, which are initiated by exciting an electron from the HOMO-4 to LUMO+4 state at time t = 0, corresponding to the ultraviolet-light excitation in experiments. [34] An example of the time evolution of the eigenenergies is shown in Fig. 7.5. Just after the excitation, the wave function of the occupied LUMO+4 is distributed mainly in the left periphery. At about 6 fs, a transition from LUMO+4 to LUMO+3 occurs, accompanied by the transfer of the electron to the right periphery. Note that the eigenenergy of the right periphery is not always lower than that of the left periphery due to their crossings. At about 17 fs, another transition to LUMO+2 occurs, causing the wave function of the occupied state to reside mainly within the core. The process of this transition is shown in the right panel of Fig. 7.5.

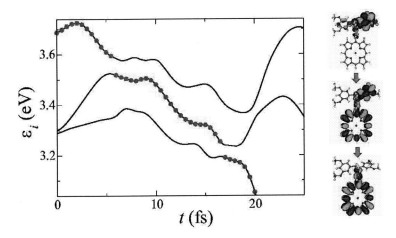

Figure 7.5. (Left panel) Time evolution of electronic eigenenergies in TDKS-FSSH simulation. The red circles denote energies of the electronic states occupied by the photoexcited electron. (Right panel) The process of the transition at about 17 fs.

7.4.4. Transfer Rate of a Photoexcited Electron

Additional TDKS-FSSH simulations are carried out using the model shown in Fig. 7.3 to estimate the electron transfer time. Figure 7.6 shows the time evolution of the existence probability $R_{core}(t)$ of a photoexcited electron in the core region obtained from the ensemble average over 15 simulations. The electron transfer time is estimated to be $\sim$ 40 fs for the current model, which suggests that it has to be measured with a resolution of this time scale.

7.5. Summary

We have investigated the dynamic properties of liquid B_2O_3 under pressures up to about 100 GPa and the energy transport mechanism in light-harvesting dendrimers by *ab initio* molecular-dynamics simulations. In liquid B_2O_3, the

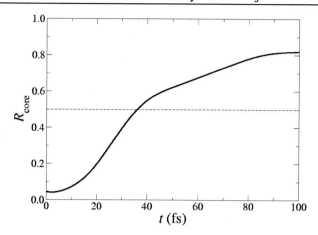

Figure 7.6. Time evolution of the existence probability $R_{core}(t)$ of a photoexcited electron in the core region. .

pressure dependence of diffusivity has been explained successfully in terms of diffusion mechanisms. For the dendrimer system, we reveal the key role of thermal molecular motion that significantly accelerate the energy transport based on the Dexter mechanism.

Acknowledgments

The present work was supported by Grant-in-Aid for JPSJ Fellows. The authors thank Research Institute for Information Technology, Kyushu University for the use of facilities. The computation was also carried out using the computer facilities at the Supercomputer Center, Institute for Solid State Physics, University of Tokyo.

References

[1] S. K. Sharma, D. Virgo, and I. Kushiro, "Relationship between density, viscosity and structure of GeO$_2$ melts at low and high pressure". *J. Non-Cryst. Solids* **33**, (1979) 235

[2] K. Funakoshi, A. Suzuki, and H. Terasaki, "*In situ* viscosity measurements of albite melt under high pressure". *J. Phys.: Condens. Matter* **14**, **(2002)**, 11343

[3] V. V. Brazhkin, I. Farnan, K. Funakoshi, M. Kanzaki, Y. Katayama, K. Trachenko, A. G. Lyapin, and H. Saitoh, "Structural Transformations and Anomalous Viscosity in the B$_2$O$_3$ Melt under High Pressure". *Phys. Rev. Lett.* **105**, (2010), 115701.

[4] N. S. Lewis, "Toward Cost-Effective Solar Energy Use". *Science* **315**, (2007) 798.

[5] T. Muraoka, K. Kinbara, and T. Aida, "Mechanical twisting of a guest by a photoresponsive host". *Nature* **440**, (2006) 512.

[6] I. Akai *et al.*, "Rapid energy transfer in light-harvesting small dendrimers". *J. Lumin.* **112**, (2005), 449.

[7] A. Yamada *et al.*, "Energy Transfer Dynamics in Light-Harvesting Small Dendrimers Studied by Time-Frequency Two-Dimensional Imaging Spectroscopy". *J. Lumin.* **129**, (2009), 1898.

[8] I. Akai *et al.*, "Rapid energy transfer in a dendrimer having p-conjugated light-harvesting antennas". *New J. Phys.* **10**, (2008), 125024.

[9] R. Kopelman *et al.*, "Spectroscopic Evidence for Excitonic Localization in Fractal Antenna Supermolecules". *Phys. Rev. Lett.* **78**, (1997), 1239.

[10] S. Raychaudhuri *et al.*, "Excitonic Funneling in Extended Dendrimers with Nonlinear and Random Potentials" *Phys. Rev. Lett.* **85**, (2000), 282.

[11] K. Nishioka, and M. Suzuki, "Dynamics of unidirectional exciton migration to the molecular periphery in a photoexcited compact dendrimer". *J. Chem. Phys.* **122**, (2005), 024708.

[12] R. Kishi *et al.*, "Core molecule dependence of energy migration in phenylacetylene nanostar dendrimers: *Ab initio* molecular orbital-configuration interaction based quantum master equation study". *J. Chem. Phys.* **128**, (2008), 244306.

[13] Y. Kodama, S. Ishii, and K. Ohno, "Dynamics simulation of a -conjugated light-harvesting dendrimer II: phenylene-based dendrimer (phDG2)". *J. Phys.: Condens. Matter* **21**, (2009), 064217.

[14] F. Vögtle, G. Richardt, and N. Werner, Dendrimer Chemistry (Wiley-VCH, Weinheim, 2009).

[15] A. Adronov, and J. M. J. Frechet, "Light-harvesting dendrimers". *Chem. Commun.* **18**, (2000), 1701.

[16] S. Ohmura and F. Shimojo, "Anomalous pressure dependence of self-diffusion in liquid B_2O_3: An *ab initio* molecular dynamics study". *Phys. Rev. B* **80**, (2009), 0202020(R).

[17] S. Ohmura and F. Shimojo, "*Ab initio* molecular-dynamics study of structural, bonding, and dynamic properties of liquid B_2O_3 under pressure". *Phys. Rev. B* **81**, (2010), 014208.

[18] P. E. Blöchl, "Projector augmented-wave method". *Phys. Rev. B* **50**, (1994), 17953.

[19] G. Kresse and D. Joubert, "From ultrasoft pseudopotentials to the projector augmented-wave method". *Phys. Rev. B* **59**, (1999), 1758.

[20] J. P. Perdew, K. Burke and M. Ernzerhof, "Generalized Gradient Approximation Made Simple". Phys. Rev. Lett. **77**, (1996), 3865.

[21] M. Tuckerman, B. J. Berne and G. J. Martyna, "Reversible multiple time scale molecular dynamics". *J. Chem. Phys.* **97**, (1992), 1990.

[22] G. J. Martyna, D. J. Tobias, and M. L. Klein, "Constant-pressure molecular-dynamics algorithms". *J. Chem. Phys.* **101**, (1994), 4177.

[23] S. Nosé, "A molecular-dynamics method forsimulations in the canonical ensemble". *Mol. Phys.* **52**, (1984), 255; W. G. Hoover, "Canonical dynamics: Equilibrium phase-space distributions" *Phys. Rev. A* **31**, (1985), 1695.

[24] O. H. Nielsen and R. M. Martin, "Quantum-mechanical theory of stress and force". *Phys. Rev. B* **32**, (1985), 3780.

[25] A. D. Corso and R. Resta, "Density-functional theory of macroscopic stress: Gradient-corrected calculations for crystalline Se". *Phys. Rev. B* **50**, (1994), 4327.

[26] S. Ohmura and F. Shimojo, "Mechanism of atomic diffusion in liquid B_2O_3: An *ab initio* molecular dynamics study". *Phys. Rev. B* **78**, (2008), 224206.

[27] J. C. Tully, "Molecular dynamics with electronic transitions". *J. Chem. Phys.* **93**, (1990), 1061.

[28] W. R. Duncan, C. F. Craig, and O. V. Prezhdo, "Time-domain *ab initio* study of charge relaxation and recombination in dye-sensitized TiO_2". *J. Am. Chem. Soc.* **129**, (2007), 8528.

[29] V. V. Brazhkin and A. G. Lyapin, "High-pressure phase transformations in liquids and amorphous solids". *J. Phys.: Condens. Matter* **15**, (2003), 6059.

[30] B. B. Karki, D. Bhattarai, and L. Stixrude, "First-principles simulations of liquid silica: Structural and dynamical behavior at high pressure". *Phys. Rev. B* **76**, (2007), 104205.

[31] V. V. Hoang, H. Zung, and N. Trung Hai, "Diffusion and dynamical heterogeneity in simulated liquid SiO_2 under high pressure". *J. Phys.: Condens. Matter* **19**, (2007), 116104.

[32] R. S. Mulliken, "Electronic Population Analysis on LCAO-MO Molecular Wave Functions. I". *J. Chem. Phys.* **23**, (1955), 1841.

[33] F. Shimojo, A. Nakano, R. K. Kalia, and P. Vashishta,"Electronic processes in fast thermite chemical reactions: A first-principles molecular dynamics study". *Phys. Rev. E* **77**, (2008), 066103.

[34] I. Akai *et al.*, "Depression of excitonic energy transfer by freezing molecular vibrations in meta-linked branching dendrimers". *Phys. Status Solidi C* **3**, (2006), 3420.

In: Molecular Dynamics of Nanobiostructures   ISBN: 978-1-61324-320-6
Editor: K. Kholmurodov   © 2012 Nova Science Publishers, Inc.

Chapter 8

MECHANIC AND ELECTRIC PROPERTIES OF GRAPHENE RIBBON-CARBON NANOTUBE NANOSTRUCTURES

Anastasia A. Artyukh[1], Leonid A. Chernozatonskii[1], Vasilii I. Artyukhov[1] and Pavel B. Sorokin[1,2,3,]*
[1]Emanuel Institute of Biochemical Physics,
Russian Academy of Sciences,
4 Kosigina st., Moscow, 119334, Russia
[2]Department of Mechanical Engineering &
Material Science and Department of Chemistry,
Rice University, Houston, Texas 77251, USA
[3]Siberian Federal University,
79 Svobodny av., Krasnoyarsk, 660041 Russia

---

[*]E-mail address: anastasiia2000@mail.ru

Abstract

This paper reviews the experimental and theoretical works on composite materials based on carbon nanotubes and graphene. Simultaneous use of 1D and 2D nanoparticles facilitates the linking between components, improving the mechanical and conductive properties of the resulting composite films in comparison with pure components. Such materials are promising for diverse applications including transparent electrodes.

Keywords: Carbon nanotubes, Graphene, Mechanic and electric properties

8.1.   Introduction

The field of polymeric nanocomposites is an area of research that has been developing intensely over the last 25 years, and encompasses diverse materials. CNT/graphene composites in question are technologically attractive because of their unique physical properties. It has been demonstrated that the presence of CNTs or graphene in composites improves the conduction and mechanical properties of source materials. Since their first discovery, carbon nanotubes (CNTs) [1] and graphene [2] have been viewed as promising materials for nanoelectronics applications, owing to their unique mechanical and electronic properties [3]. Graphene is a monolayer of sp^2-hybridized carbon atoms comprising a hexagonal structure. It is a semimetal with a zero energy gap and a so-called Dirac spectrum. Single-walled carbon nanotubes (SWCNTs) can be viewed as graphene sheets rolled up in cylinders, and they can have semiconductor or metallic properties depending on the parameters of rolling. Nanotubes can be either single- or multiwall. Multiwall CNTs (MWCNTs) have certain advantages over single-wall tubes, such as their superior mechanical stability and rigidity essential for scanning probe applications. MWCNTs are better suited for preparation of composite materials, because they can endure stress better and without damage to interior tubes [1]. Numerous nanodevices based on CNTs and graphene have been proposed, such as ultrasensitive molecule sensors [4], electron emitters based on CNTs [5, 6] or graphene [7], hydrogen storage devices [8], or field effect transistors [2]. In 2008, a composite from MWCNTs and graphene sheets was synthesized for the first time [9]. Simultaneously, simulations of covalently and molecular-bound SWCNT/graphene sheet structures were carried out [10]. Over the last two years, a diversity of experimental and

theoretical works devoted to such structures came out. It was demonstrated that
the properties of these materials surpass the individual mechanical and electronic properties of CNTs or graphene. Such materials show great promise for
future applications [11].

8.2. Fabrications of Composites

In 2008, several structural models of composites containing only CNTs and
single-layer graphene were proposed [10]. It was noted that, since graphene
and CNTs can be produced in solution, and their edges are chemically active,
they should be expected to form mixed composites with different energetically
favorable configurations [10, 15, 16, 17]. First, CNTs can be attached through
their open ends to the plane of graphene sheet (Figure 4, a). There are two different cases: a - nanotube "grown" from graphene with sp2 hybridization atoms
on interface, b - nanotube attached to graphene with sp3 bonds on interface. For
instance, as in case of Figure 4-a,b, transition region contains a 6 heptagons.
The third and fourth structure types are formed when the nanotube lies horizontally on the graphene sheet, with carbon atoms at the interface assuming sp3
hybridization. In this case, the structure of nanotube ends plays the major role:
if the ends are open, the CNT and graphene attract and form covalent sp3 bonds
(Figure 6, b), whereas if the ends are capped, the two components of the system
are bound by intermolecular forces, but no covalent bonds form between them.
The final possibility is when the edge of graphene layer is joined to a generator
line of the CNT, with sp3-hybrid atoms at the interface (Figure 4, c).

For this time several methods of preparing carbon nanotube/graphene composites were developed. A greater part of these methods based on mechanical
mixing of individual soluble components, with subsequent sedimentation and
drying. At the same time, other interesting methods have been proposed employing self-assembly or even direct CVD growth of the composite material.

In 2008, researchers from the Institute of Polymer Technology and Materials Engineering (UK) fabricated hybrid films composed of graphite oxide and
MWCNTs [12]. For this, graphite oxide was prepared and purified according
to the above Hummers technique[13]. Individual graphite oxide sheets were
treated by ultrasound at room temperature. MWCNTs with hydroxyl groups
were dissolved in (N,N-dimethyl)formamide under unltrasonication at room
temperature. Individual graphite oxide sheets and dissolved SWCNTs were
mixed, coated on a glass surface and dried for 1 d at 100^0. Thickness of the

resulting film was controlled by the volume of mixture put on the substrate, and varied from 2 to 8 μm. It was noted that the main factor governing the resistivity of the material was the percentage of MWCNTs in the composite. The resistivity decreases with lessening of percentage of MWCNTs and with film thickness. The structure of resulting films is illustrated in Figure 8.1. In 2009, another method was proposed to fabricate similar films composed of two layers (graphite oxide and MWCNT) via self-assembly [14]. For this, graphite oxide prepared via the Hummers technique [13] was exfoliated and dispersed in distilled water using ultrasound treatment. The solution was centrifuged for 10 min. For film fabrication, an aminated SiO2/Si substrate was dipped in the as-prepared solution, washed with water and ethanol, and dried in nitrogen flow. After that the substrate was immersed in water solution of aminated MWCNTs. The substrate was subsequently washed with water and ethanol, dried in nitrogen flow, and subjected to thermal treatment at 125^0 for 15 min. The scheme of preparation is presented in Figure 8.1.

After this, films were deoxidized in a hydrazine monohydrate solution at different (N,N-dimethyl)formamide concentrations for 1 d at 80^0 and annealed at 500^0 in argon atmosphere. As a result, a large percentage of graphene sheets was formed in the film, which increased their conductivity by orders of magnitude. According to atomic force and scanning electron microscopy images, the as-produced films are homogeneous. They exhibit mechanical strength, which enable a long service time for these thin conducting transparent films.

A purely-carbon structure composed of nanotubes and graphene sheets was produced for the first time using chemical vapor deposition [9]. For this, catalyst-cobalt and titanium nitride film-was sputtered onto a silicon substrate. Then, the substrate was heated in a low-pressure chamber. Argon-acetylene mixture at a 1000 Pa pressure was used as a carbon source. The temperature of the substrate was 510^0, growth was carried out for 10 min, and the rate of growth was about 670 nm/min. The resulting composite structure is shown in Figure 8.3. It was characterized using scanning electron microscopy (SEM) and transmission electron microscopy. The width of graphene sheets varies from 17 to 38 nm and depends on the width of the sputtered catalyst layer, specifically, the cobalt film (the titanium nitride film width was fixed at the same 5 nm value throughout the experiments). The spacing between graphene layers is 0.36 nm. All resulting nanotubes turned out to be multiwalled with an average diameter of 11.9 nm. The as-prepared composite has a flat surface and produces good electrical and thermal contacts with other materials in all directions. Therefore,

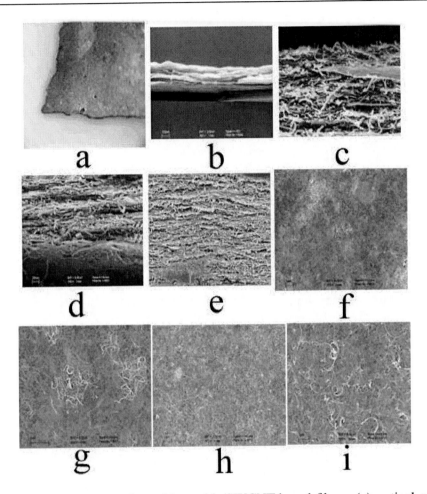

Figure 8.1. Structure of graphite oxide/SWCNT-based films: (a) optical microscopy image; (b) SEM images of transverse film cross-sections; (c) SEM image of 2-?m thick film cross-section, the ratio of graphite oxide:SWCNTs is 1:0.5; (d) SEM image of 3.5-?m thick film cross-section, the ratio of graphite oxide:SWCNTs is 1:0.5; (e) SEM image of 8-?m thick film cross-section, the ratio of graphite oxide:SWCNTs is 1:0.5; (f, g) surface images of films, the ratio of graphite oxide:SWCNTs is 1:0.5; (h, i) surface images of films, the ratio of graphite oxide:SWCNTs is 1:5 [12]

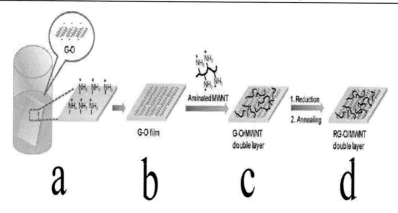

Figure 8.2. Preparation of two-layer films: (a) an aminated SiO2/Si substrate is immersed in aqueous solution containing graphite oxide (G-O); (b) the substrate with a deposited G-O film is (c) coated by aminated nanotubes; after (c) deoxidation and annealing, (d) a double RG-O/MWCNT layer is formed [13]

its further applications in electronics and other fields is anticipated.

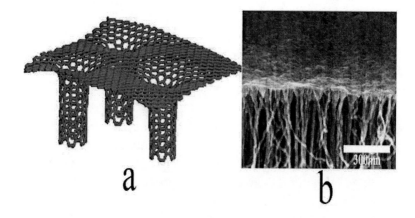

Figure 8.3. (a) A schematic illustrating the attachment of CNTs to graphene; (b) SEM image of the as-fabricated composite [9].

8.3.   Theoretical Studies

8.3.1.   Structure

In 2008, several structural models of composites containing only CNTs and single-layer graphene were proposed [10]. It was noted that, since graphene and CNTs can be produced in solution, and their edges are chemically active, they should be expected to form mixed composites with different energetically favorable configurations [10, 15, 16, 17]. First, CNTs can be attached through their open ends to the plane of graphene sheet (Figure 8.4, a). There are two different cases: a - nanotube "grown" from graphene with sp^2 hybridization atoms on interface, b - nanotube attached to graphene with sp^3 bonds on interface. For instance, as in case of Figure 8.3-a,b, transition region contains a 6 heptagons. The third and fourth structure types are formed when the nanotube lies horizontally on the graphene sheet, with carbon atoms at the interface assuming sp3 hybridization. In this case, the structure of nanotube ends plays the major role: if the ends are open, the CNT and graphene attract and form covalent sp^3 bonds (Figure 8.6,b), whereas if the ends are capped, the two components of the system are bound by intermolecular forces, but no covalent bonds form between them. The final possibility is when the edge of graphene layer is joined to a generator line of the CNT, with sp3-hybrid atoms at the interface (Figure 8.4,c).

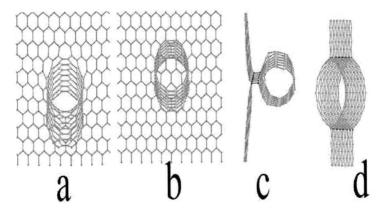

Figure 8.4. Structural models of composites: (a, b) open-end attachment of the CNT (6,6) to graphene sheet; (c) parallel CNT (5,5) attachment to graphite; (d) attachment of the graphene sheet to CNT (12,0) wall [10].

Such a configurations can form because the most chemically active atoms of a nanographene (NG) are those at its edges [10, 18]. Figure 8.5 presents a scheme of attachment of a graphene nanofragment to a CNT based on molecular dynamics and quantum chemistry calculations. As the edge of the nanographene approaches the CNT fragment, they are attracted to each other, forming a line of sp3-hybrid atoms. In the end, the shape of the CNT fragment changes from cylindrical to oval (Figure 8.5,c). Using the (8,0) nanotube as an example, it was shown that C atoms on the opposing side of the CNT become more chemically active than the rest. In solution, it is favorable for a second NG sheet to approach and attach to the CNT-NG structure from the opposite side of the interface. Another possibility is when an incoming nanotube fragment attaches to the chemically NG edge of the structure, forming a CNT-NG-CNT structure [14]. Attachment of several (2, 4, 6) NGs to SWCNT wall has also been investigated-see Figure 8.6. The binding energy E of composite structures was estimated using the following formula:

$$(E = E_{comp} - E_{CNT} - nE_g)/N_{atoms} \qquad (8.3.1)$$

where E_{comp} is the energy of the composite cluster, E_{CNT} is the energy of the carbon nanotube, E_g is the energy of the graphene ribbon, and n is the number of graphene fragments attached to the nanotube. Natoms is the total number of atoms in the unit cell. It was found that the binding energy of zigzag CNT with two zigzag graphene nanoribbons (CNT/2ZGNR) (Figure 8.6,a) is 0.0151 eV/atom, while the addition of two fragments is energetically favorable with a binding energy of 0.0244 eV/atom for CNT/4ZGNR (see Figure 8.6,b) and of 0.1900 eV/atom for CNT/6ZGNR (see Figure 8.6,c). These results demonstrate the stability of the proposed structures.

In solution-phase CNT/NG mixtures, formation of supramolecular structures is also possible. Using the molecular dynamics method, the process of molecular "capture" of NG on the surface of a nanotube was investigated [18]. Figure 8.7 shows the progression of supramolecular attachment of a square-shaped NG fragment to a SWCNT. It was demonstrated that if the shape of the NG is a ribbon of width comparable to CNT diameter, in the optimized structure, the nanoribbon adheres to the nanotube throughout its length at $0 - 300°K$.

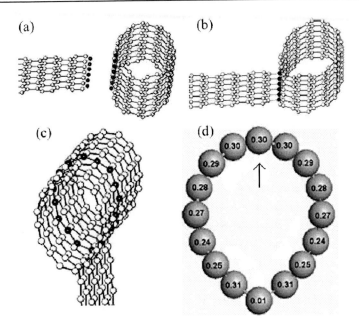

Figure 8.5. Covalently bonded structures: (a) initial configuration prior to attachment of the nanographene (NG) to an armchair CNT (6,6); (b) optimized structure upon attachment; (c) an optimized NG-zigzag CNT (8 ,0) structure with highlighted atoms across the CNT circumference; (d) chemical activity of the highlighted atoms of CNT (8, 0) (numbers represent the unpaired electron density, the arrow points at the most active atom) [18].

8.3.2. Electronic Properties

The electronic properties of SWCNTs with side-attached graphene sheets were also studied [17]. The electronic spectrum of such a structure can be qualitatively represented as the superposition of the spectra of the graphene fragment and an SWCNT with hydrogen atoms attached along the attachment line.

The electronic structure of the whole composite is not simply the superposition of bands of the NT and graphene strip (GS). The local densities of states of the GS and NT parts are shown in Figure 8.8 (a) in green and red lines respectively. As seen from the figure, some Van Hove peaks correspond to the nanotube and some other ones correspond to graphene. But the local DOS of the nanotube bears little resemblance to the DOS of an isolated (12,0) NT be-

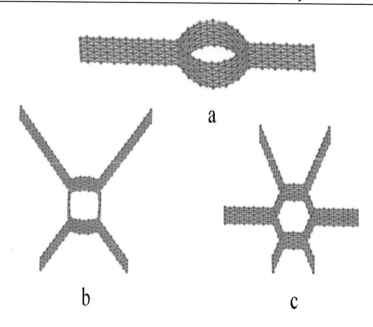

Figure 8.6. Structure of zigzag graphene nanoribbon and (12,0) carbon nanotube composites: (a) two-leafed CNT/2ZGNR, (b) four-leafed CNT/4ZGNR, and (c) six-leafed CNT/6ZGNR [18].

cause of the chemical attachment of graphene significantly alters the atomic structure of the nanotube, and therefore, modifies the electronic structure of the NT (a similar result was seen in the case of adsorption of Al and single hydrogen atoms [19]). The local DOS of the NT has a lot of common features with the DOS of the NT with adsorbed hydrogen atom pair. The common peaks are depicted by vertical gray lines. It should be mentioned that GS/NT system and the NT with adsorbed hydrogen atom pair have a similar band gap (0.08 eV).

Similar changes in the electronic spectrum occur when two graphene nanostrips are attached to the flanks of the CNT (Figure 8.9) [17]. The total spectrum of the structure corresponds qualitatively to the superposition of GNS and isolated CNT spectra with shape change induced by addition of hydrogen atoms along two contact lines. It has been demonstrated that the attachment of graphene fragments to the nanotube does not result in a loss of conductivity of the composite. It can be concluded that CNT/2GNS composite display metallic

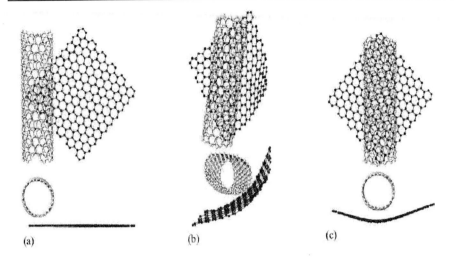

Figure 8.7. Process of supramolecular structure formation from a square NG and a CNT from starting (a) structure to (c) optimal configuration [18].

properties. The part of the structure responsible for electric conductance is the deformed CNT.

8.3.3. Mechanical Properties

It was demonstrated that in CNTs with side-attached graphene sheets, addition of more sheets ¿results in higher Young's module [16]. Thus, the calculated Young's modulus of a SWCNT is 1.34 TPa, which is in accordance with experimental data (these range from 0.40 to 4.15 TPa). The Young's moduli of structures with 2, 4, and 6 sheets are 2.77, 2.83, and 3.04 TPa, respectively. We supposed, that this increase take place due to enhancing a number of sp^3 atom interfaces. The behavior of the structures upon bending and twisting was studied, as well (Figure 8.10). It was found that upon bending, structures with 2 and 4 GNRs attached to the nanotube convert to another stable state via the deformation of nanotube and the formation of additional sp3 bonds on a bent. The new state is energetically more favorable than the starting unbent state, and the latter is only restored upon heating.

Figure 8.10 shows that at 2 and 4 degree bending angles, metastable states are observed, having energies in excess of that of the initial structure.

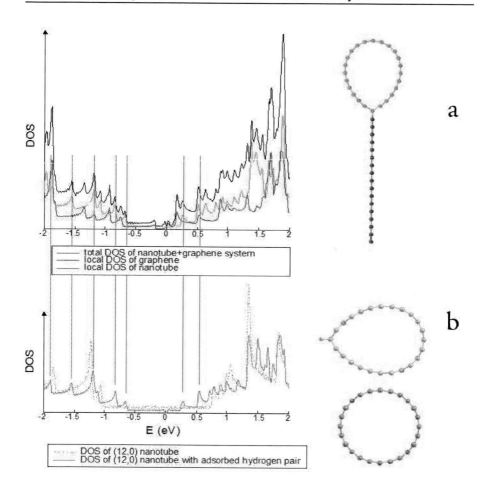

Figure 8.8. (a) Total and local density of states of the composite GS/NT structure. The total DOS is shown in black, the local DOS of the NT part is shown in green, and the local DOS of the GS part is shown in red. The GS/NT structure is shown on the right with parts colored in correspondence with the local DOS plots. (b) Total DOS of the (12,0) and total DOS of the (12,0) nanotube with adsorbed hydrogen atom pair. The energy zero is at the Fermi level. (c, d) Attachment of an armchair GS edge to the (12,0)NT/GS nanotube (starting and final configurations) [17].

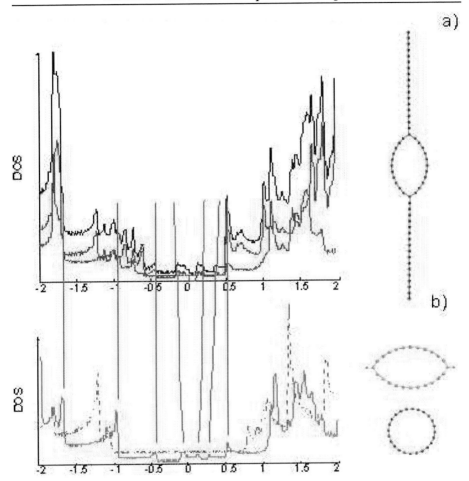

Figure 8.9. (a) The local DOS of the CNT/2ZGNR composite. The total density of states is represented by the black line, the local DOS of the CNT part is shown in light gray line, the local DOS of the GNR part is presented as gray line. The CNT/2ZGNR structure is presented on the right-hand part of the picture, the color coding CNT and ZGNR parts corresponds to the local DOS plots; (b) The total DOS of the (12,0) CNT is presented as dashed line, the total DOS of the (12,0) CNT with adsorbed hydrogen atom pairs is presented as solid line. The zero of energy is taken at the Fermi level.

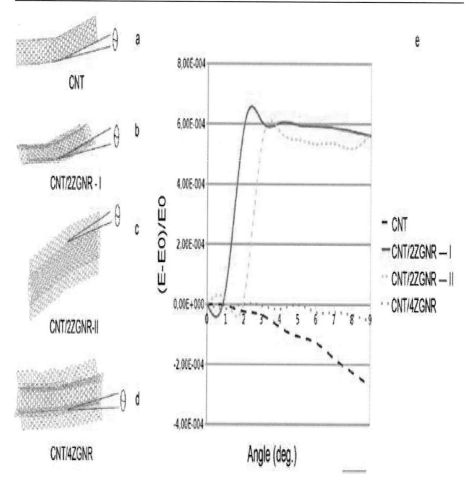

Figure 8.10. (a) (12,0) CNT, (b, c) CNT/2GNR and (d) CNT/4GNR structures with indicated directions of bending; (e) reduced deformation energies as a function of the bending angle [16].

8.4. Conclusions

In summary, there are several techniques of preparation of carbon nanotube-graphene composites. The properties of the resulting composite films show improvement (higher mechanical strength, lower resistivity) over respective pure components, and depend crucially on the optimal balance between the percentage of 1D and 2D nanoparticles. Though most of the experiments performed to date used multiwalled carbon nanotubes, single-walled nanotubes have also been successfully mixed with graphene to produce composite materials. Theoretical studies confirm that single-walled carbon nanotubes covalently linked with graphene sheets improve the conductivity and mechanical properties of the composites compared to pure parent materials. Such materials show great promise for fabrication of conductive films, and such composites are likely to find their applications in various nanodevices and nanoelectromechanical systems.

References

[1] M.S. Dresselhaus, G. Dresselhaus, P. Avouris (Eds.), "Carbon nanotubes: synthesis, structure, properties, and applications". *Topics in Applied Physics Series, Springer, Berlin*, Vol. 80, (2000).

[2] K.S. Novoselov, A.K. Geim, S.V. Morozov, D. Jiang, Y.Zhang, S.V. Dubonos, I.V., Grigorieva, A.A. Firsov, "Electric Field Effect in Atomically Thin Carbon Films". *Science*, **306**, (2004), 5696: 666-669.

[3] C. Berger, Z.M. Song, X.B. Li, X.S. Wu, N. Brown, C. Naud, D. Mayo, T.B. Li, J. Hass, A.N. Marchenkov, E.H. Conrad, P.N. First, W.A. de Heer, "Electronic Confinement and Coherence in Patterned Epitaxial Graphene". *Science*, **312**, (2006), 1191.

[4] F. Schedin, A.K. Geim, S.V. Morozov, E.W. Hill, P. Blake, M.I. Katsnelson, K.S. Novoselov, "Detection of individual gas molecules adsorbed on graphene". *Nat. Mater.*, **6**, (2007), 652.

[5] L.A. Chernozatonskii, Yu.V. Gulyaev, Z.Ja. Kosakovskaja, N.I. Sinitsyn, G.V. Torgashov, Yu.F. Zakharchenko, E.A. Fedorov and V.P. Val'chuk, " Electron field emission from nanofilament carbon films". *Chemical Physics Letters*, **233**, (1995), 63-68.

[6] S.M. Jung, J. Hahn, H.Y. Jung, and J.S. Suh, *Nano Lett.*, **6**, (2006), 7.

[7] Z.-S. Wu, S. Pei, W. Ren, D. Tang, L. Gao, B. Liu, F. Li, C. Liu, and H.-M. Cheng, "Clean Carbon Nanotube Field Emitters Aligned Horizontally". *Adv. Mater.*, **21**, (2009), 1756-1760.

[8] A.C. Dillon, K.M. Jones, T.A. Bekkedahl, C.H. Kiang, D.S. Bethune, M.J. Heben, "Storage of hydrogen in single-walled carbon nanotubes". *Nature*, **386**, (1997), 377-379.

[9] D. Kondo, S. Sato, and Y. Awano, "Self-organization of Novel Carbon Composite Structure: Graphene Multi-Layers Combined Perpendicularly with Aligned Carbon Nanotubes". *Applied Physics Express*, **1**, (2008), 074003: 16.

[10] E.F. Sheka, L.A. Chernozatonskii, "Graphene-carbon nanotube composites". *MOLMAT2008, Toulouse, France, July 8-11th 2008, Abstracts, P26; arXiv: 0901.3624v1, http://arxiv.org/abs/0901.3624 (22.01.2009)*, (2008).

[11] J.K. Wassei, R.B. Kaner, "Graphene, a promising transparent conductor". *Materials today*, **13**, (2010), 3.

[12] D. Cai, M. Song, C. Xu, "Highly Conductive Carbon-Nanotube/Graphite-Oxide Hybrid Films". *Adv. Mater.*, **20**, (2008), 1706-1709.

[13] W.S. Hummers, R.E. Offeman, "Preparation of Graphitic Oxide". *J. Am. Chem. Soc.*, **80**, (1958), 1339.

[14] Y.-K. Kim, D.-H. Min, "Durable Large-Area Thin Films of Graphene/Carbon Nanotube Double Layers as a Transparent Electrode". *Langmuir*, **25**, (2009), 19: 11302-11306.

[15] E.F. Sheka, L.A. Chernozatonskii, "Graphene-Carbon Nanotube Composites". *J. Comput. Theor. Sci.*, **27** (2010), 9: 1814.

[16] A.A. Artyukh, L.A. Chernozatonskii, and P.B. Sorokin, "Mechanical and electronic properties of carbon nanotube-graphene compounds". *Phys. Status Solidi B*, **247**, (2010), 11-12: 2927-2930.

[17] L. Chernozatonskii, P. Sorokin, "Graphene-nanotube Structures: Architecture, Properties and Applications". *ECS Trans.*, **19**, (2009), 13.

[18] L.A. Chernozatonskii, E.F. Sheka, A.A. Artyukh, "Graphene-nanotube structures: Constitution and formation energy". *JETP Letters*, **89**, (2009), 7: 352-356.

[19] O. Gülseren, T. Yildirim, and S. Ciraci, "Tunable Adsorption on Carbon Nanotubes". *Phys. Rev. Lett.*, **87**, (2001), 116802.

In: Molecular Dynamics of Nanobiostructures ISBN: 978-1-61324-320-6
Editor: K. Kholmurodov © 2012 Nova Science Publishers, Inc.

Chapter 9

MODELING STRUCTURE OF THE LIGHT HARVESTING COMPLEX LH1 FROM THE BACTERIAL PHOTOSYNTHETIC CENTER OF *Thermochromatium tepidum*

Bella L. Grigorenko[1], *Maria G. Khrenova*[1],
Alexander V. Nemukhin[1,2] *and Jian P. Zhang*[3]*

[1]Department of Chemistry, M. V. Lomonosov Moscow State
University, Moscow, Russian Federation
[2]N. M. Emanuel Institute of Biochemical Physics, Russian
Academy of Sciences Moscow, Russian Federation
[3]Department of Chemistry, Renmin University of China,
Beijing, China

---

*E-mail address: anemukhin@yahoo.com

Abstract

Structure and function of the light harvesting complex LH1 of the bacterial photosynthetic center of a thermophilic purple sulfur bacterium, *Thermochromatium tepidum*, present a challenge in studies of photoreceptor biomolecular systems since the experimental spectroscopy information is not currently augmented by the available crystallography data. By using the primary sequence of amino acid residues of the α- and β-polypeptide helices from this LH1 complex and the related templates we constructed the three-dimensional structural model of the entire antenna system. Molecular dynamics simulations of this system solvated in water both for calcium free and calcium bound units show that the calcium ions can be trapped at least at two different binding sites. Quantum chemical calculations of the bacteriochlorophyll absorption bands at the predicted LH1 geometry configurations indicate that one of these sites is preferable to explain the observed red shifted absorption upon calcium binding.

Keywords: Photosynthesis, Light harvesting complexes, Quantum calculations

9.1.   Introduction

Light harvesting (LH) complexes are the important components of photosynthetic centers in plants and bacteria. These biomolecular units capture the sun light and transfer the energy in the form of electronic excitation to the reaction centers. Structure and function of the LH1 complex from the thermophilic purple sulfur bacterium, *Thermochromatium tepidum* [1], present a challenge in studies of these transmembrane proteins since the experimental spectroscopy information [2–6] is not currently augmented by the available detailed crystallography data. The primary sequence of amino acid residues of the α- and β-polypeptide helices surrounding bacteriochlorophyll (BChl) cofactors from this LH1 complex is known [5], but information on the three-dimensional molecular structure has been lacking. A number of high-resolution (2.0 - 2.5 Å) crystal structures are available for another type LH complex, LH2, from purple bacteria *Rhodopseudomonas acidophila* [7–9] or *Rhodospirillum molischianum* [10], while the structures obtained for the LH1 complex at a low resolution 4.8 Å [11] may pretend only for a general view. According to these data [11] the LH1 complex consists of 16 subunits, i.e. the pairs of the α- and β -polypeptide helices capping the BChl-a dimers.

An interest to the light harvesting complex from *Thermochromatium tepidum* first of all is due to the thermophylic properties of this photosynthetic bacterium which can grow at high temperatures up to 58°C. The LH1 complex is tightly associated with the reaction center (RC), and the extracted from the organism the LH1-RC unit is also a thermostable species [2].The spectral studies of LH1-RC show that this complex is characterized by an unusual red shift of the absorption band ascribed to the Qy band of BChl-a from 885 nm in the mesophilic counterparts to 915 nm in LH1-RC [2–4]. Kimura and co-authors have shown that the calcium ions are responsible both for the enhanced thermal stability and for the unusual red shift of LH1-RC from *Thermochromatium tepidum* [5,6]. Another important observation is that each of 16 subunits of the and α- and β-polypeptide pairs contains one calcium ion.

Previously two attempts have been performed to restore the full-atom three-dimensional structures of the pair of the α- and β-polypeptide helices surrounding the BChl-a dimer in the LH1 complex from *Thermochromatium tepidum* by using molecular modeling tools. Ma et al. [4] applied the SWISS-MODEL and MODELER algorithms and used the α, β-polypeptide units from the LH2 complex from *Rhodopseudomonas acidophila* [9] and *Rhodospirillum molischianum* [10] as the templates. The authors tentatively ascribed the calcium ion binding sites to the loop domains at the C-terminus of the helices. Later, Grigorenko et al. [12] constructed this α-, β-helices, BChl-a dimer subunit by manually aligning the structures with the help of the template from the LH2 complex of *Rhodopseudomonas acidophila* [9]. The constructed structures were refined by using the molecular mechanics algorithms. Another binding site for the Ca^{2+} ions was suggested as that located in the immediate vicinity of the BChl-a dimer. In both works no attempts have been carried out to estimate the absorption bands.

Here we suggest a full-atom three-dimensional model for the entire LH1 complex with all 16 subunits as constructed by template simulations and refined in molecular dynamics (MD) calculations. According to the present MD simulations for the fully solvated systems with calcium free and calcium bound units the calcium ions can be trapped at least at two different binding sites tentatively suggested previously. Quantum chemical calculations of the bacteriochlorophyll absorption bands at the predicted LH1 geometry configurations indicate that one of these sites is preferable to explain the observed red shifted absorption upon calcium binding.

9.2. Modeling and Results

Fig 9.1 illustrates the composition of each subunit which includes the peptide chains of the α- and β-helices with the given sequences of amino acid residues. Two parallel solid lines indicate the BChl-a dimer showing the magnesium ions located inside the corresponding rings. Coupling of the α- and β-parts occurs at the level of the α-His34 and β-His36 residues coordinated by magnesium ions.

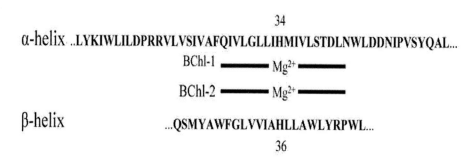

Figure 9.1. Composition of each subunit of the LH1 complex from *Thermochromatium tepidum.*

Fig 9.2 shows the view of the coupling region at the atomic resolution; for better visibility we only indicate the heavy atoms.

Following the previous strategy [12], construction of each subunit (Fig. 9.1, Fig. 9.2) was performed by using the template from the LH2 complex of *Rhodopseudomonas acidophila* [9]. The structural fragments were aligned by the positions of the α-His34 and β-His36 residues and continued manually in both directions from histidines with a partial molecular mechanics optimization at each step. Next, the entire LH1 system composed of 16 subunits was created and fully solvated by water molecules. In total, more than 250000 atoms in the model system were taken into account in the MD simulations. Calculations of MD trajectories were performed using the NAMD 2.6 software suite [13] freely available at http://www.ks.uiuc.edu/Research/namd/. The CHARMM force field parameters [14] for protein atoms and Ca^{2+}, and the TIP3P model parameters for all water molecules were employed. CHARMM general force field parameters [15] were used for BChl-a. The MD simulations were carried out with a 1 fs integration step following initial 20000 step energy minimization. The system was gradually heated during 300 ps to 300K, 500 ps

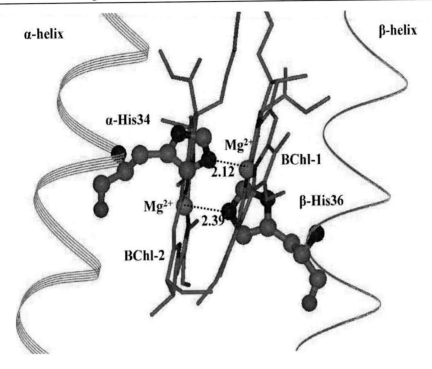

Figure 9.2. Fragment of the α-, β-, BChl-a dimer subunit showing the heavy atoms only. Carbon atoms are colored in green, nitrogen in blue, oxygen in red, magnesium in magenta. Distances between Mg^{2+} ions and Nε atoms from His are given in angstroms.

at 300K and cooled down to 0 K during 1500 ps with the subsequent 5000 step minimization. No restrictions were imposed on coordinates of all atoms in trajectory calculations. The VMD program [16] was used for visualization (http://www.ks.uiuc.edu/Research/vmd/).

Firstly, the calcium free system was considered. Fig. 9.3 shows the general view of the model system after removal of all water molecules. The unconstrained energy minimization allowed us to obtain three-dimensional all-atom equilibrium geometry parameters for the LH1 complex without calcium ions. Next, 16 calcium ions were added to the model system following previous two [4, 12] suggestions. In both cases the MD protocol applied for the calcium free system was repeated. We observed that Ca^{2+} ions remained trapped inside the protein either at the C-terminus [4] or near the BChl dimer [12] along the

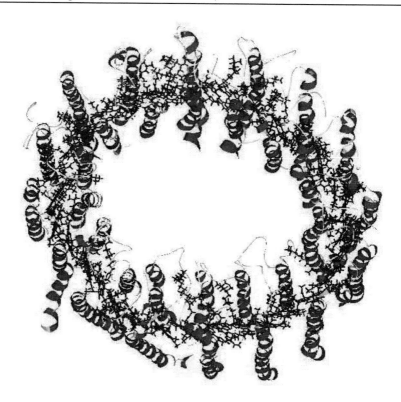

Figure 9.3. Structure of the LH1 complex from *Thermochromatium tepidum* constructed by template and refined in MD simulations. For better visibility, the α- and β-helices (in ribbon representation) and BChl-a dimers (in sticks) are only shown.

trajectories. Energy minimization for both calcium bound species resulted in the corresponding equilibrium geometry configurations. We paid attention to the equilibrium geometry parameters of all 32 BChl-a molecules.

To estimate positions of the absorption band maxima of the calcium free and calcium bound systems we considered the BChl-a molecules at the corresponding equilibrium geometry configurations inside the LH1 complex. Therefore we assume that possible shifts in the absorption spectra due to capture of calcium ions by the protein may be, at least, partially ascribed to the conformational changes upon metal binding. We considered every BChl-a molecule from the set of 32 species at three respective protein conformations, namely, without calcium ions and with calcium ions at two possible binding sites. A series of 96 quantum chemical calculations was performed for the vertical excitation energies of BChl-a in the ZINDO approximation [17]. Although the wavelengths of the absorption band maximum predicted by such a procedure lacks a sufficient accuracy (the values around 1000 nm have been obtained), the tendencies in the band shifts should be of significance. From our calculations we conclude that the calcium binding site at the C-terminus is consistent with the observed red shift, while the site near BChl-a is responsible for mainly blue shifts in absorption.

9.3. Conclusion

By the motifs of the primary sequence of amino acid residues of the α- and β-polypeptide helices of the LH1 complex from *Thermochromatium tepidum* we constructed the three-dimensional structural model of the entire antenna system. As shown in large-scale MD simulations of this system solvated in water both for calcium free and calcium bound units show that the calcium ions can be trapped at least at two different binding sites. Estimates of the absorption band maxima performed with the ZINDO quantum chemical method allowed us to conclude that the calcium binding site at the C-terminus is consistent with the observed red shifted absorption upon calcium binding.

Acknowledgments

The work was supported by the Russian Foundation for Basic Researches (project No. 08-03-92203-GFEN-a). We thank the Research Computing Center of the M.V. Lomonosov Moscow State University for providing computational facilities.

References

[1] Madigan M.T.,"A novel photosynthetic purple bacterium isolated from a yellowstone hot spring". *Science*, **225**, (1984), 313-315.

[2] Suzuki, H., Hirano, Y, Kimura, Y., Takaichi, S., Kobayashi, M., Miki, K., Wang Z.Y.,"Purification, characterization and crystallization of the core complex from thermophilic purple sulfur bacterium Thermochromatium tepidum". *Biochim. Biophys. Acta*, **1767**, (2007), 1057-1063.

[3] Ma, F., Kimura, Y., Zhao, X.H., Wu, Y.S., Wang, P., Fu, L.M., Wang, Z.Y., Zhang J.P., "Purification, characterization and crystallization of the core complex from thermophilic purple sulfur bacterium Thermochromatium tepidumExcitation dynamics of two spectral forms of the core complexes from photosynthetic bacterium *Thermochromatium tepidum*". *Biophys. J.*, **95**, (2008), 3349-3357.

[4] Ma, F., Kimura, Y., Yu, L.J., Wang, P., Ai, X.C., Wang, Z.Y., Zhang J.-P., "Specific Ca^{2+}-binding motif in the LH1 complex from photosynthetic bacterium *Thermochromatium tepidum* as revealed by optical spectroscopy and structural modeling". *FEBS J.*, **276**, (2009), 1739-1749.

[5] Kimura. Y., Hirano, Y., Yu, L.J., Suzuki, H., Kobayashi, M., Wang Z.Y., "Calcium ions are involved in the unusual red shift of the light-harvesting 1 Qy transition of the core complex in thermophilic purple sulfur bacterium *Thermochromatium tepidum*". *J. Biol. Chem.*, **283**, (2008), 13867-13873.

[6] Kimura. Y., Yu, L.J., Hirano, Y., Suzuki, H., Wang Z.Y., "Calcium ions are required for the enhanced thermal stability of the light-harvesting-reaction center core complex from thermophilic purple sulfur bacterium *Thermochromatium tepidum*". *J. Biol. Chem.*, **284**, (2009), 93-99.

[7] McDermott, G., Prince, S.M., Freer, A.A., Hawthornthwaite-Lawless, A.M., Papiz, M.Z., Cogdell, R.J., Isaacs N.W., "Crystal structure of an integral membrane light-harvesting complex from photosynthetic bacteria". *Nature*, **374**, (1995), 517-521.

[8] Papiz, M.Z., Prince, S.M., Howard, T., Cogdell R.J., Isaacs N.W. "The structure and thermal motion of the B800-850 LH2 complex from *Rps.acidophila* at 2.0 Åresolution and 100K: new structural features and functionally relevant motions". *J. Mol. Biol.*, **326**, (2003), 1523-1538.

[9] Cherezov V., Clogston J., Papiz M.Z., Caffrey M., "Room to move: crystallizing membrane proteins in swollen lipidic mesophases". *J. Mol. Biol.*, **357**, (2006), 1605-1618.

[10] Koepke, J., Hu, X., Muenke, C., Schulten, K., Michel H., "The crystal structure of the light-harvesting complex II (B800-850) from *Rhodospirillum molischianum*". *Structure*, **4**,(1996), 581-597.

[11] Roszak, A.W., Howard, T.D., Southall, J., Gardiner, A.T., Law, C.J., Isaacs, N.W., Cogdell R.J.,"Crystal structure of the RC-LH1 core complex from Rhodopseudomonas palustris". *Science*, **302**, (2003), 1969-1972.

[12] Grigorenko, B.L., Nemukhin, A.V., Zhang, J.P., Wang P., "Modeling calcium binding at the light harvesting complex of the bacterial photosynthetic center of *Thermochromatium tepidum*". *Moscow State Univ. Res. Bull., Chemistry*, (2010), in press.

[13] Phillips, J.C., Braun, R., Wang, W., Gumbart, J., Tajkhorshid, E., Villa, E., Chipot, C., Skeel, R.D., Kale, L., Schulten K., "Scalable molecular dynamics with NAMD". *J. Comp. Chem.*, **26**, (2005), 1781-1802.

[14] MacKerell, Jr., A.D., Bashford, D., Bellott, M., Dunbrack Jr., R.L., Evanseck, J.D., Field, M.J., Fischer, S., Gao, J., Guo, H., Ha, S., Joseph-McCarthy, D., Kuchnir, L., Kuczera, K., Lau, F.T.K., Mattos, C., Michnick, S., Ngo, T., Nguyen, D.T., Prodhom, B., Reiher, III, W.E., Roux, B., Schlenkrich, M., Smith, J.C., Stote, R., Straub, J., Watanabe, M., Wiorkiewicz-Kuczera, J., Yin, D., Karplus M., " All-atom empirical potential for molecular modeling and dynamics studies of proteins". *J. Phys. Chem. B*, **102**, (1998), 3586-3616.

[15] Vanommeslaeghe, K., Hatcher, E., Acharya, C., Kundu, S., Zhong, S., Shim, J., Darian, E., Guvench, O., Lopes, P., Vorobyov, I., MacKerell Jr. A.D., "CHARMM General Force Field (CGenFF): "A force field for drug-like molecules compatible with the CHARMM all-atom additive biological force fields". *J. Comp. Chem.*, **31**, (2010), 671-690.

[16] Humphrey, W., Dalke, A., Schulten K., "VMD: visual molecular dynamics". *J. Molec. Graphics.*, **14**, (2006), 33-38.

[17] Zerner M.C., in *Rev.Comput.Chem.*, Ed. K.B. Lipkowitz and D.B. Boyd, Vol.2 (VCH Publishing, New York, 1991), 313-366.

In: Molecular Dynamics of Nanobiostructures   ISBN: 978-1-61324-320-6
Editor: K. Kholmurodov   © 2012 Nova Science Publishers, Inc.

Chapter 10

MOLECULAR DYNAMICS SIMULATION OF WATER ADSORBED ON ICE NUCLEATION PROTEIN

Daisuke Murakami and Kenji Yasuoka[*]
Center for Applied and Computational Mechanics
School of Science for Open and Environmental Systems
Keio University, Yokohama, Japan

Abstract

An ice nucleation protein induces phase transition from liquid water to ice in the air. A specific hydrophilic surface of the protein may have an influence on the network of hydrogen bonds between water molecules adsorbing onto the protein. However, microscopic characteristics of the ice nucleation protein and the behavior of water molecules on the protein have not been clarified. Therefore, molecular dynamics simulations of a system consisting of water and an ice nucleation protein was used to clarify some dynamics in the atomic level. As a result, there were some differences between simulation predictions of water clusters adsorbed on the ice nucleation protein and the conventional percolation theory. It was found that finite clusters tend to be localized on the surface and trapped

---

[*]E-mail address: daisukem@a2.keio.jp

by sites of the protein. The initial results suggested the need for study on another type of hydrophilic protein and weaker hydrophilicity. The results pointed out the fact that the hydrophilicity of the ice nucleation protein influenced the formation of the water network that water clusters adsorbed on the ice nucelation protein tend to be localized.

Keywords: Molecular dynamics simulation, Water, Ice nucleation protein

Introduction

Ice nucleation proteins exist on outer cell membranes of an ice nucleation active bacteria. This bacteria has been said to cause frost damage which means that organizations of leaves of crops are damaged by ice in the air. This bacteria has been considered to play an important role in phase transitions of liquid water to ice in the air, and the ice nucleation protein has been considered to be the main source of the phenomenon. The applications are expected to be in industrial technologies for ice thermal storage or production of artificial snow. However the 3D structure of the ice nucleation protein is difficult observing by X-ray diffraction. Only some amino acid sequences were clarified. Moreover, the mechanism causing frost damage is unclear.

Some models of the 3D structure of the ice nucleation protein have been proposed. inaZ [1] is a model of ice nucleation proteins. inaZ includes some experimental parameters regarding the ice nucleation protein. The most characteristic aspect is that there is a large repetitive central domain. The primary structure of the ice nucleation protein repeats itself with a high fidelity of 48 residues and consists of beta-conformations. Considering those facts, inaZ makes Ser, Thr and Asp residues forming zigzag structures exposed in the air. Those formations seemed to be a random ice Ic template. Therefore moisture in the air is predicted to freeze around the ice. We used this model as an ice nucleation model and performed molecular dynamics simulations to obtain knowledge of the phase transition.

The research process starts by gathering knowledge of the transition from liquid water to ice on the ice nucleation protein. To understand the microscopic behavior of water molecules influenced by the ice nucleation protein, we performed classical molecular dynamics simulations. Specifically, we constructed a system of quasi-2D water clusters covering an ice nucleation protein and analyzed the dynamics. We changed the number densities of the water molecules that form the quasi-2D water clusters on the ice nucleation protein and compared results with the conventional percolation theory and another type of hydrophilic protein. We also weakened values of charges of the ice nucleation protein and studied the influences of the hydrophilicity of the protein.

Method and Theory

Figure 10.1 - 10.3 is the procedure on how to make the initial structure of the simulated system. Firstly, we made a system of *inaZ* and equilibrated the system to conduct MD simulations (Fig. 10.1). Secondly, we selected *inaZ* and water molecules within 0.5 nm from the surface of the protein, and we replicated the system four times (Fig. 10.2). Lastly, we eliminated some water molecules at random to make quasi-2D water clusters with various densities covering *inaZ* (Fig. 10.3). The position of *inaZ* was fixed during all simulations, and water molecules only had interactions with *inaZ* and other water molecules.

Table 10.1 is a summary of the simulation conditions. We performed identical simulations for each density, where the temperature was fixed at 300 K. Simulations were run for 500 ps to obtain initial configurations. Data for analysis were sampled during the last 100 ps, and MDGRAPE-3 which is a special purpose computer for classical molecular dynamics simulations [2] was used.

Table 10.1. Simulation Condition

INP model	*inaZ*
Water model	TIP4P
Number of atoms of INP	12,648
Number of water molecules	240 - 1920
Boundary Condition	Periodic
Ensemble	NVT
Temperature	300 K
Time scale	500 ps
Temperature contorolling method	velocity scaling
Time integration method	Symplectic integration
Force Field	AMBER96
Computer	PC + MDGRAPE-3

Figure 10.1. The first step of the simulation procedure. A simulation was done to relax a system of *inaZ* and water. Blue atoms are water molecules.

The percolation theory [3] deals with the science of connection and exhibits universality. For example, we arrange a 2D square lattice model (Fig. 10.4). When two neighboring lattice points are occupied, we assume that they are connected. If we change an occupation probability of total lattice points in the system from 0 to 1, an infinite spanning cluster emerges at some probability (Fig.10.5). That probability is the so-called percolation threshold (Fig. 10.6). Some physical values near the threshold diverge and that behavior depend only on dimensions, not structure. For instance, the mean cluster size, S_{mean}, is subject to Eqn.(10.0.1), (Fig. 10.7). s is the number of the particles within the cluster and n_s is the number of s-size clusters in the system. C_c^* is the threshold of the density and γ is a critical exponent that obey depends on the dimension, not on the structure. So we derived physical values of water molecules to compare the results of the ice nucleation protein with the one of the percolation theory and another type of protein.

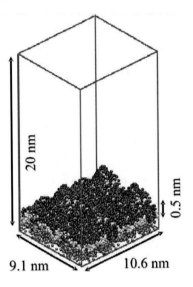

Figure 10.2. Nextly, we chose water molecules within 0.5nm from the surface of *inaZ*.

$$S_{\text{mean}} \equiv \sum \frac{n_s s^2}{\sum n_s s} = |\, C^* - C_c^* \,|^{-\gamma} \qquad (10.0.1)$$

Results and Discussion

Figure 10.8 and 10.9 illustrate snapshots of water clusters of different densities. The former never percolates, but the latter always does. But at an intermediate density, two cluster states, spanning or non-spanning, could be observed. In Fig. 10.10 we plotted those probabilities via C^*. The line is an approximation obtained by least squares fitting of the data by

$$R(C^*) = 0.50 + 0.50 \tanh (C^* - a)/b \qquad (10.0.2)$$

where a and b are fitting parameters.

We also derived S_{mean} in Fig. 10.11 that excluded the spanning clusters. Behavior of those two physical values are different from those of the percolation theory. In the percolation theory, slopes of the two values should diverge at the

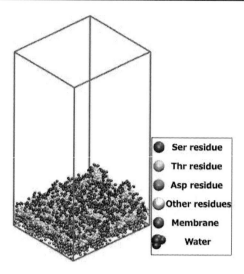

Figure 10.3. Lastly, we eliminated water molecules at random. After that, we conducted simulations with various densities of water molecules.

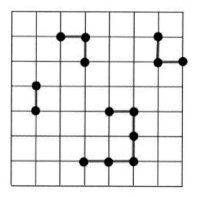

Figure 10.4. There are only finite clusters at a low occupation probability (Black points and blue lines represent finite clusters).

threshold. The percolation probability should take either zero or one and the mean cluster size should increase exponentially near the percolation threshold. However, the results of R and S_{mean} nearby C^* do not behave as above.

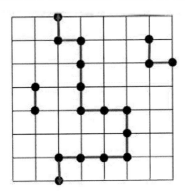

Figure 10.5. There are both spanning and non spanning clusters at a high occupation probability (Red lines represent a spanning cluster).

We derived two water densities by using Eqn. (10.0.2), $C_1^* = 5.3$ nm^{-2} at $R = 0.50$ and $C_2^* = 6.8$ nm^{-2} at $R = 0.99$, to compare another hydrophilic protein (lysozyme [4]) and table 10.2 represents those results. It was found that both densities around the percolation thresholds were nearly the same, but the slant of the percolation probability of *inaZ* was smoother than that of lysozyme, which was estimated by those two densities. The results also showed that finite water clusters are captured more locally by hydrophilic sites on the ice nucleation protein surface than lysozyme.

Table 10.2. Comparison of two hydrophilic proteins

Density	INP	Lysozyme [4]
C_1^* nm^{-2}	5.3	5.8
C_2^* nm^{-2}	6.8	6.5

Simulations in the case of weak charges were also conducted to understand the hydrophilicity of the surface of the ice nucleation protein.

Figure 10.12 and 10.13 show formations of water clusters near the percolation threshold. Although the densities of the two figures were the same, there were many differences in the water behavior. There were much more local finite clusters with the default charge than with the half charge. Water clusters were

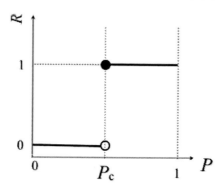

Figure 10.6. The occupation spanning probability. In the percolation theory, the spanning probability takes either 0 or 1.

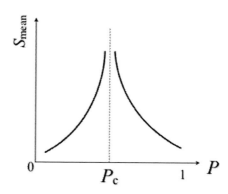

Figure 10.7. The mean cluster size via the occupation probability. In theory, S_{mean} should behave as Eqn. 10.0.1.

formed dispersively and larger local finite clusters remained. On the other hand from Fig. 10.13, it seemed that local finite water clusters tend to be captured by infinite clusters. There also seemed to be a particular area where water clusters were more likely to form.

Figure 10.14 is the percolation probability via the number of quasi-2D density of water molecules and Fig. 10.15 shows the mean finite cluster sizes. In both analysis, slants of physical values of the half charge tend to be more di-

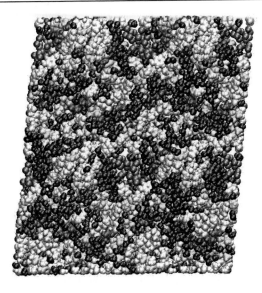

Figure 10.8. A snapshot at a low density of water.

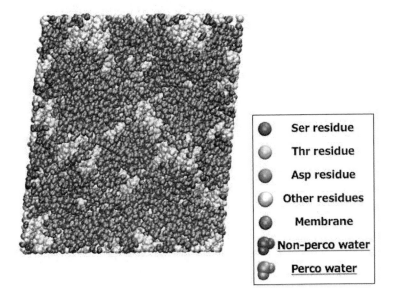

Figure 10.9. A snapshot at a high density of water.

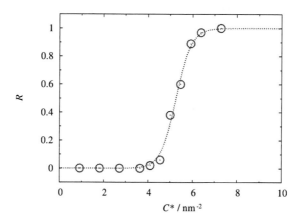

Figure 10.10. The spanning probability via the number of quasi-2D density of water on *inaZ*.

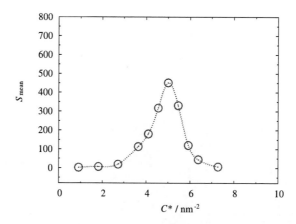

Figure 10.11. The mean cluster size via the number of quasi-2D density of water on *inaZ*.

vergent near the percolation threshold than those of the default charge. It was found that water behavior on a weakened hydrophilicic surface was closer to the conventional percolation theory.

Figure 10.12. A snapshot in the case of the default charge.

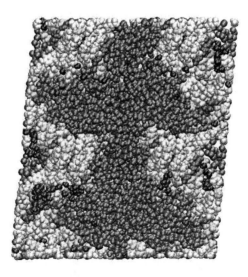

Figure 10.13. A snapshot in the case of the half charge. The density is the same as the one in Fig. 10.12.

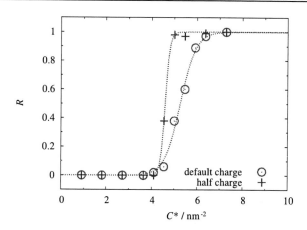

Figure 10.14. A comparison of the spanning probalility between the default charge and half charge.

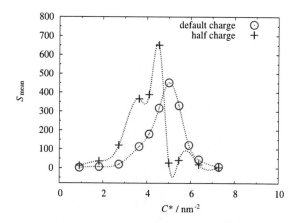

Figure 10.15. A comparison of the mean cluster size between the default charge and half charge.

Conclusion

We studied molecular dynamics of water adsorbed on an ice nucleation protein. This protein has been said to cause frost damage, but the microscopic mechanism and characteristics of the ice nucleation protein have not been clarified. Therefore we performed simulations on a system for the ice nucleation protein and quasi-2D water clusters on it to understand the phase transition process. As a result, it was found that water clusters on the ice nucleation protein tend to be localized, because the water clusters were inconsistent with the conventional percolation theory near the threshold. The same tendency also occurs from comparisons with another protein type and an ice nucleation protein with weakened hydrophilicity. As a consequence, the ice nucleation protein influenced water clusters that were likely to be localized and showed some different behavior than from the bulk state. In the future, we will try to perform several simulations with a long time scale or a lower temperature where water is likely to be in an amorphous state to analyze dynamics of water in detail and retrieve more knowledge about the phase transition.

References

[1] Kajava, A.V. and Lindow, S.E., "A Model of the 3-Dimensional Structure of Ice Nucleation Proteins". *J. Mol. Biol.*, **232(3)**, (1993), 709-717.

[2] Narumi, T., "A 55 TFLOPS simulation of amyloid-forming peptides from yeast prion sup35 with special-purpose computer system MDGRAPE-3". In: *Proc. Supercomputing*, (2006).

[3] In book: Stauffer D. and Aharony A., "Introduction to Percolation Theory". *Taylor and Frances Ltd, 11 New Fetter Lane, London EC4P 4EE*, (1994).
See also URL http://www.taylorandfrancis.com/.

[4] Oleinikova A. et al., "Formation of Spanning Water Networks on Protein Surfaces via 2D Percolation Transition". *J. Phys. Chem. B*, **109(5)**, (2005), 1988-1998.

In: Molecular Dynamics of Nanobiostructures   ISBN: 978-1-61324-320-6
Editor: K. Kholmurodov   © 2012 Nova Science Publishers, Inc.

Chapter 11

MD SIMULATIONS ON THE STRUCTURE OF ONCO-PROTEINS P53: WILD-TYPE AND RADIORESISTANT MUTANT SYSTEMS

Kholmirzo T. Kholmurodov[1,2], *Evgenii A. Krasavin*[1,2],
Viktor A. Krylov[1], *Ermuhammad B. Dushanov*[1],
Vladimir V. Korenkov[1,2], *Kenji Yasuoka*[3], *Tetsu Narumi*[4],
Yousuke Ohno[5], *Makoto Taiji*[5] *and Toshikazu Ebisuzaki*[6]*

[1]Joint Institute for Nuclear Research (JINR), Dubna,
Moscow Region, Russia
[2]Dubna International University,
Dubna, Moscow Region, Russia
[3]Keio University, Yokohama, Japan
[4]University of Electro-Communications, Tokyo, Japan
[5]RIKEN-Yokohama Institute, Yokohama, Japan
[6]The Institute of Physical and Chemical Research (RIKEN),
Saitama, Japan

*E-mail address: mirzo@jinr.ru

Abstract

Based on molecular dynamics (MD) simulation, a comparative analysis has been performed of the p53 dimer – DNA interaction for the wild-type and mutant Arg273His (R273H) proteins. The aim of this paper is to study the molecular mechanism of the p53 onco-protein and DNA binding. A comparative analysis shows that the R273H mutation has a significant effect on the p53–DNA interaction removing their close contact. The obtained MD simulation results illustrate in detail the molecular mechanism of the conformations of the key amino acids in the p53–DNA binding domain, which is important for the physiological functioning of the p53 protein and understanding the origin of cancer.

Keywords: MD simulations, Onco-protein p53, p53R273H mutation

11.1. Introduction

p53 (the 53 kilodalton (kDa) protein, also known as protein 53 or tumor protein 53) is activated either to induce a cell cycle arrest allowing the repair and survival of the cell, or apoptosis to discard the damaged cell. The p53 tumor suppressor protein is involved in preventing cancer and plays a central role in conserving genomic stability by preventing a genome mutation. When activated, p53 binds DNA and activates the expression of several genes including WAF1/CIP1 encoding for p21. p21 (WAF1) binds to the G1-S/CDK (CDK2) and S/CDK complexes (molecules important for the G1/S transition in the cell cycle) inhibiting their activity [1-5].

In Fig.1, the side and top views of the p53 mouse protein are presented (PDB entry file: 3EXJ). For the p53 dimer structure, two chains (A and B) which surround DNA symmetrically are shown along with two catalytic centers [Zn(CYS)$_3$(HIS)$_1$] (the zinc-binding interfaces are displayed in the right pictures; zinc is shown as a grey sphere).

Single amino acid substitutions (mutations) in the p53 structure deactivate the p53 protein, which results in cancer [1-8]. Usually, most mutations (95% of all known tumor mutations) occur in the DNA-binding domain (DBD) of the p53 protein. Thus, an oncogenic form of p53 is predominantly a full-length p53 protein with a single amino acid substitution in the DBD. Most of these mutations destroy the ability of the protein to bind to its target DNA sequences, and thus prevent the transcriptional activation of these genes. Tumors with inactive p53 mutants are aggressive and often resistant to ionizing radiation and chemotherapy.

Some reported observations with regard to the p53 mutations could be outlined as follows [1-8]. The effects of the G245S, R248Q, R249S, and R273H mutations disturb essentially the stability of the p53 core domain. A detailed analysis of p53 mutations shows that the vast majority of the mutations of p53 cluster in conserved regions of the DNA-binding core domain (CD) (residues 96-292). Around 20% of all mutations are concentrated at five "hotspot" codons in the core domain: 175, 245, 248, 249, and 273.

In [4], examining the Arg273His (R273H) mutation and p53–DNA interactions, it was found that at least three R273H monomers are needed to disable the p53 tetramer, which is consistent with experiments:
—a single R273H monomer may greatly shift the binding mode probabilities;
—the work suggests that p53 needs balanced binding modes to maintain genome stability.

In [6], it has been shown that mutations in p53 occur at a rate of approximately 70% in hormone-refractory prostate cancer (CaP). The observations are summarized as follows:
—the R273H p53 mutation (p53R273H) facilitates androgen-independent (AI) growth in castrated nude mice;
—p53R273H-mediated AI CaP can induce the expression of prostate-specific antigen via an androgen receptor-mediated pathway.

In [7], it has been established that p53 is the most commonly mutated tumour suppressor gene in human cancers. The most common p53 cancer mutations (R248W and R273H) gain novel oncogenic activities. The important observations allow concluding that:
—the tumor supressor functions of p53 are abolished in p53-mutant mice;
—p53 gain-of-function mutants promote tumorigenesis by a novel mechanism involving active disruption of critical DNA damage-response pathways;
—translocations, a type of genetic instability rarely observed in p53-/- cells, are readily detectable in p53-mutant pre-tumor thymocytes.

As regards simulation, molecular dynamics (MD) is one of the powerful techniques to study the p53 protein structural behavior and its mutation transition. Nevertheless, few published papers addressed p53 protein simulation based on the MD approach [9-11]. Two major computational difficulties that block the efficient use of MD for the p53 protein could be outlined (as detailed

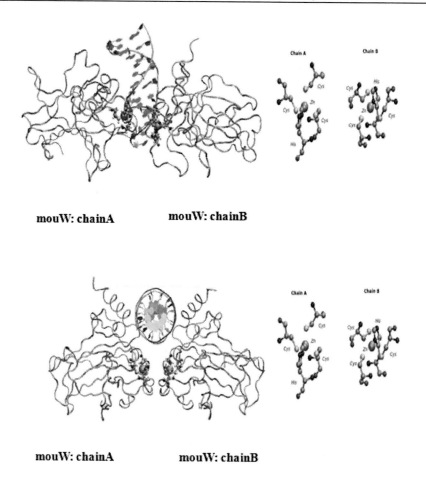

mouW: chainA   mouW: chainB

mouW: chainA   mouW: chainB

Figure 11.1. The side and top views of the p53 mouse protein are shown (PDB entry file: 3EXJ). For the p53 protein structure, two DNA chains (A and B) are shown along with two catalytic centers [Zn(CYS)$_3$(HIS)$_1$] of the zinc-binding interface (Zinc is shown as grey spheres).

in [9]) as follows: (1) a large size of p53; (2) the problem of accurate modeling the zinc-binding interface in p53c. The p53 protein functions as a tetramer in a cell, but the p53c monomer is already 50 Å in diameter. Simulation of the p53 tetramer structure in explicit water has to be extremely difficult because a large number of water molecules are needed to solvate this protein. The MD simulations performed in [9] were limited to the consideration of the p53c monomer only. It was found that the monomer p53c model is stable in vitro and that such model is still suitable for the protein stability analysis, although the monomer p53c–DNA binding is weaker.

In the present study, we have focused on the p53 dimer (the A and B chains) structure to investigate the p53–DNA binding phenomena. The presence of the p53 dimer structure is a more adequate model in comparison to the monomer p53c ones. In the dimer representation, the p53 protein will symmetrically surround the relevant DNA molecule from two sides, which allows the formation of a binding interface in a native manner. Next, we perform a comparative MD analysis for the p53 dimer – DNA interaction between the wild-type and mutant Arg273His (R273H) version of the p53 protein. Finally, apart from the previous MD studies [9-11], we investigate the correlation effects between two different, but structurally identical, models - the p53 human and mouse proteins.

11.2.   Mouse p53 Structure: The Effect of the R273H Mutation on the p53–DNA Binding Domain

Four structures have been simulated using the MD method in the same environment (a water solvent) and simulation conditions (temperature, pressure, and the ensemble):
—(1 and 2) The wild-type (mouW) and mutant R273H (mouM) mouse p53 proteins,
—(3 and 4) The wild-type (humW) and mutant R273H (humM) human p53 proteins.

In total, we have performed 16 model calculations (periodic PME-NPT; periodic PME-NVT; non-periodic cutoff; and non-periodic no cutoff) on four p53-relevant structures (the mouse p53 protein: a wild-type and mutant R273H version (PDB entry file "3EXJ"); the human p53 protein: a wild-type and mutant R273H version (PDB entry file "1TSR")) – to compare the conformational changes between the relaxed and original 3D states in the native and mutant structures (see Appendix for the MD simulation details).

In Fig.2, the side and top views of the p53 mouse protein are shown (PDB entry file: 3EXJ). For the p53 protein structure, two chains (A and B) symmet-

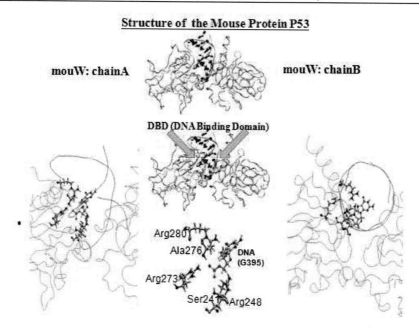

Figure 11.2. The side and top views of the p53 mouse protein are shown (PDB entry file: 3EXJ). For the p53 protein structure two chains (A and B) are symmetrically surround (yellow arrows) the related DNA sequence located in the central domain DBD (DNA binding domain).

rically surround (yellow arrows) the related DNA sequence located in the central DNA binding domain (DBD). In the p53-DBD interaction, three arginines (R248, R273, and R280), one serine (S241), and one alanine (A276) are responsible for DNA binding. In Fig.2, a positional snapshot of the p53 chain A {Arg248, Arg273, Arg280, Ser241, Ala276} with respect to DNA is separately displayed.

Considering the above noted difficulty of zinc binding in p53c, experiments point to the importance of Zinc coordination for achieving the correct folding and correct binding of p53 to a specific DNA in intact cells. In simulation, however, both non-bonded and bonded approach are in use to describe the zinc-p53c binding interface. It should be stressed that it is quite challenging to maintain the interface stability at room temperature during long-time simulations [9]. We have followed this criteria and have not fixed zinc locally in its binding interface, even though it is suitable to simulate proteins in which zinc is required for the catalytic function [9]. In this study, we focus on the peculiarities of global structural changes of the p53 protein dimer.

We have performed a comparative analysis of the wild-type and mutant version between the mouse (mouW, mouM) and human (humW, humM) p53 proteins. For both mouse and human p53 structures, we have examined the effect of the Arg273His (R273H) mutation on the p53–DNA binding domain.

Arginine (Arg) is an α-amino acid with $pK_a=12.48$; its side chain consists of a 3-carbon aliphatic straight chain and is positively charged in neutral, acidic, and even most basic environments. Because of conjugation between the double bond and the nitrogen lone pairs, the positive charge is delocalized, enabling the formation of multiple H-bonds. Histidine (His) is an aromatic amino acid with $pK_a=6.5$; its side chain consists of a positively charged imidazole ring which is aromatic at all pH values. This means that at physiologically relevant pH values, relatively small shifts in pH will change its average charge. Below a pH of 6, the imidazole ring is mostly protonated [15]. The differencies in the chemical structure and properties between arginine and histidine during nanosecond dynamical changes could influence the final (relaxed) states of all amino acids. In Fig.3, the position of {Arg273His} → {DNA} for the p53 mouse protein (chain A) are shown at t=0 and 3 ns.

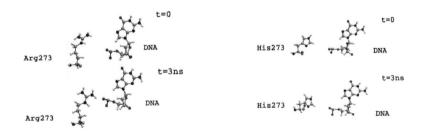

Figure 11.3. The snapshots show amino acid positions {Arg273His} → {DNA} for the p53 mouse protein (chain A; left: wild-type; right: mutant (R273H)) at t=0 and 3 ns.

The statistics below presents MD calculation results on the positional changes of five amino acid residues, which are related to the direct p53–DNA contact:

(mouW, mouM)

—(p53' chain A) Res{Arg248, Arg273His, Arg280, Ser241, Ala276} →
{DNA}
—(p53' chain B) Res{Arg391, Arg416, Arg423, Ser384, Ala419} → {DNA}
(humW, humM)
—(p53' chain A) Res{Arg248, Arg273His, Arg280, Ser241, Ala276} →
{DNA}
—(p53' chain B) Res{Arg344, Arg369, Arg376, Ser337, Ala372} → {DNA}

While constructing the p53–DNA distance diagrams below (Figs.4-13
(mouW, mouM) and Figs.16-25 (humW, humM)), we have esimated 5 distance
distributions between the DNA phosphorus atom and the corresponding residue
atom (from the N- to CO- ends):
d_1[Res(N)–DNA(P)] – blue; d_2[Res(CA)–DNA(P)] – green; d_3[Res(CB)–
DNA(P)] – light blue; d_4[Res(C)–DNA(P)] – violet; d_5[Res(O)–DNA(P)] – red.

(1) In Fig.4 (left: mouW & right: mouM), the 3-ns dynamics of the R248–
DNA (DG395) interaction distance is shown for the mouse p53 protein (chain
A). The results indicate that the R273H point mutation (p53' chain A) causes
some changes in orientation for the neighbouring residue R248. However, the
R248–DNA distance kept around an average value of 7-8 Å as for the wild type.
In Fig.5 (left: mouW & right: mouM), the 3-ns dynamics of the R344–DNA
(DT407) interaction distance is shown for the mouse p53 protein (chain B).
Like in Fig.4, when the wild-type and mutant proteins are compared, the results
point to orientational changes for the R344 residue. The R344 residue from
chain B forms a closer contact with DNA than R248–DNA from chain A of the
p53 protein (R344–DNA distance kept around an average value 3-4 Å).

(2) In Fig.6 (left: mouW & right: mouM), the 3-ns dynamics of the R273–
DNA (DG395) interaction distance is shown for the mouse p53 protein (chain
A). For the wild-type protein, the minimal R273–DNA distance is around 4 Å,
while for the mutant protein the minimal H273–DNA distance increases up to
8 Å. It is obvious that the R273H mutation removes a close contact of the p53
protein with DNA. In Fig.7 (left: mouW & right: mouM), the 3-ns dynamics
of the the R369–DNA (DT407) interaction distance is shown for the mouse
p53 protein (chain B). Comparring Figs. 6 and 7, we see that the R273H point
mutation (induced in chain A) causes a reverse effect on opposite chain B, where
the R369–DNA contact distance decreases.

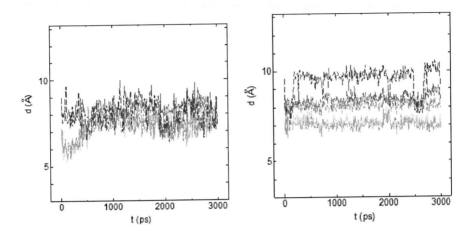

Figure 11.4. Distance diagrams of the 3-ns dynamics of the Arg248–DNA (DG395) contact of the mouse p53 protein (chain A). The wild-type p53 protein (left); the mutant (R273H) p53 protein (right).

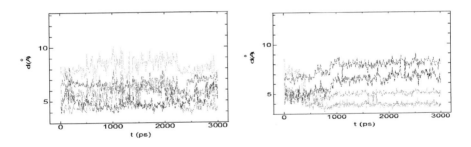

Figure 11.5. Distance diagrams of the 3-ns dynamics of the Arg344–DNA (DT407) contact of the mouse p53 protein (chain B). The wild-type p53 protein (left); the mutant (R273H) p53 protein (right).

(3) In Fig.8 (left: mouW & right: mouM), distance diagrams for the 3-ns dynamics of the R280–DNA (DG395) interaction are shown for the mouse p53 protein (chain A). In Fig.9 (left: mouW & right: mouM), distance diagrams for the 3-ns dynamics of the R376–DNA (DT407) interaction are shown for the mouse p53 protein (chain B). In Figs.8 and 9, we observe similar distance distributions between the wild-type and mutant proteins. Obviously, for both A and B chains, there are no visible effects of the R273H mutation on the R280–

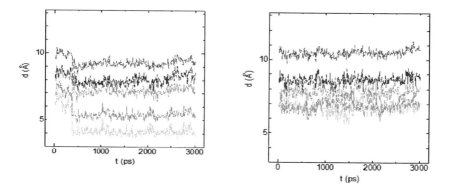

Figure 11.6. Distance diagrams of the 3-ns dynamics of the Arg273–DNA (DG395) contact of the mouse p53 protein (chain A). The wild-type p53 and mutant p53 protein (R273H) (left and right, respectively).

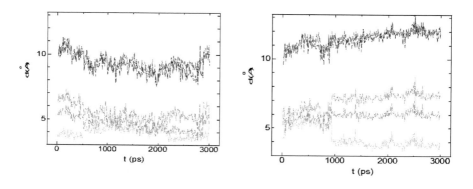

Figure 11.7. Distance diagrams of the 3-ns dynamics of the Arg369–DNA (DT407) contact of the mouse p53 protein (chain B). The wild-type p53 and mutant p53 protein (R273H) (left and right, respectively).

DNA and R376–DNA distances.

(4) In Fig.10 (left: mouW & right: mouM) The 3-ns dynamics of the S241–DNA (DG395) interaction distance is shown for the mouse p53 protein (chain A). The S241–DNA distance fot the wild-type protein slightly increases, whereas it decreases for the mutant protein. It should be stressed that in chain

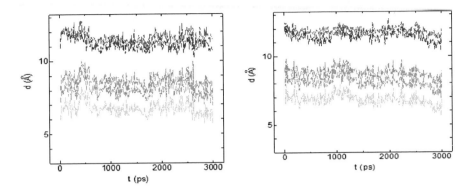

Figure 11.8. Distance diagrams of the 3-ns dynamics of the Arg280–DNA (DG395) contact of the mouse p53 protein (chain A). The wild-type p53 protein (left); the mutant (R273H) p53 protein (right).

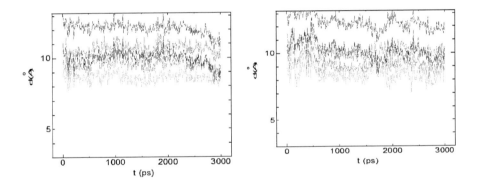

Figure 11.9. Distance diagrams of the 3-ns dynamics of the Arg376–DNA (DT407) contact of the mouse p53 protein (chain B). The wild-type p53 protein (left); the mutant (R273H) p53 protein (right).

A of p53, the S241–DNA distance has the smallest values among all five amino acid residues of the DNA binding domain. In Fig.11 (left: mouW & right: mouM), the 3-ns dynamics of the S337–DNA (DT407) interaction distance is shown for the mouse p53 protein (chain B). The S337–DNA distance of the wild-type protein decreases, whereas that of the mutant protein increases. Like in p.(2), the R273H point mutation (induced in chain A) causes a reverse effect on oppsotie chain B.

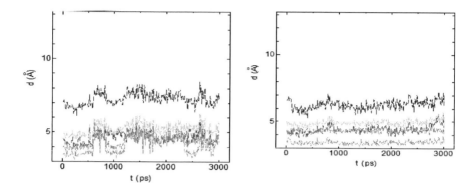

Figure 11.10. Distance diagrams of the 3-ns dynamics of the Ser241–DNA (DG395) contact of the mouse p53 protein (chain A). The wild-type protein (left); the mutant (R273H) p53 protein (right).

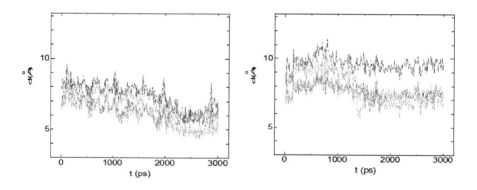

Figure 11.11. Distance diagrams of the 3-ns dynamics of the Ser337–DNA (DT407) contact of the mouse p53 protein (chain B). The wild-type protein (left); the mutant (R273H) p53 protein (right).

(5) In Fig.12 (left: mouW & right: mouM), the 3-ns dynamics of the A276–DNA (DG395) interaction distance is shown for the mouse p53 protein (chain A). At the final stages, the A276–DNA distance of the wild-type protein increases, while that of the mutant one keeps constant. In Fig.13 (left: mouW & right: mouM), the 3-ns dynamics of the A372–DNA (DT407) interaction distance is shown for the mouse p53 protein (chain B). The behavior of the A372–DNA distance is similar to the results presented in pp.(2) and (4). In chain B of

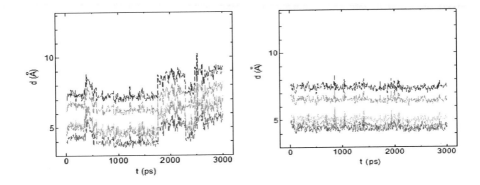

Figure 11.12. Distance diagrams of the 3-ns dynamics of the Ala276–DNA (DG395) contact of the mouse p53 protein (chain A). The wild-type protein (left); the mutant (R273H) p53 protein (right).

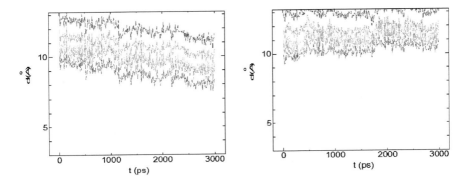

Figure 11.13. Distance diagrams of the 3-ns dynamics of the Ala372–DNA (DT407) contact of the mouse p53 protein (chain B). The wild-type protein (left); the mutant (R273H) p53 protein (right).

p53, the A372–DNA distance of the wild-type protein decreases, whereas that of the mutant protein increases. Obviously, for the A372–DNA contact point, the R273H mutation (induced in chain A) causes a reverse effect on oppsotie chain B.

The MD results for the p53–DNA distances described above (Figs.4-13) are illustrated by the snapshot below. In Fig.14, the positions of amino acids

{Arg248, Arg280, Ser241, Ala276} → {DNA} are shown for t=0 and 3 ns (chain A of the p53 mouse protein: wild-type (left); R273H mutant (right)). The comparison of the p53-DNA distances in Fig.14 is straightforward.

11.3.   Human p53 Structure: The Effect of the R273H Mutation on the p53–DNA Binding Domain

In this section, like for the mouse p53 protein, we perform MD analysis of the human p53 structure and investigate the effect of the R273H mutation on the p53–DNA binding domain. In Fig.15, the side and top views of the p53 human protein are shown (PDB entry file: 1TSR). For the p53 protein structure, two chains (A and B) symmetrically surround (yellow arrows) the related DNA sequence located in the central DNA binding domain (DBD). Below we discribe the calculation results on the positional changes of all five amino acid residues {Arg248, Arg273, Arg280, Ser241, Ala276} (chain A of p53) and {Arg391, Arg416, Arg423, Ser384, Ala419} (chain B of p53), which are related to the direct p53–DNA contact.

To compare the structural behavior of the human and mouse p53 proteins, an important peculiarity of the DNA starting conformation should be noted. Chain A of the mouse p53 protein forms a closer contact with DNA than chain B, whereas in the human p53, it is chain B that forms a closer contact with DNA. It should be stressed that, considering again the effect of the R273H mutation, the R273H amino acid is located on chain A, which is farther from DNA than chain B. So, performing analysis for the human and mouse proteins, we should compare the following data:

—{hum(W,M), chain B} → {mou(W,M), chain A}
—{hum(W,M), chain A} → {mou(W,M), chain B}

In Fig.15, a snapshot of chain B of p53 {Arg391, Arg416, Arg423, Ser384, Ala419}, which is responsible for DNA binding, is separately displayed.

The statistics below present MD calculation results on positional changes of five amino acid residues, which are related to the direct p53–DNA contact: (humW, humM)

—(p53' chain A) {Arg248, Arg273His, Arg280, Ser241, Ala276} → {DNA}
—(p53' chain B) {Arg344, Arg369, Arg376, Ser337, Ala372} → {DNA}

(1) In Fig.16 (left: humW & right: humM) the interaction distance R248–DNA (DC15) during 3-ns dynamical changes are shown for the human p53 protein (chain A). The average R248–DNA distance for both humW and humM proteins are around 4-5 Å, although for the wild-type more close contact can be observed. The results indicate on some orientation changes of residue R248 as like as for the chain B of mouse p53 (Fig.5). In Fig.17 (left: humW & right: humM) the interaction distance R391–DNA (DG13) during 3-ns dynamical changes are shown for the human p53 protein (chain B). The correlation of the results for the distance distribution between the human and mouse proteins are obvious (compare Fig.17 and Fig.4). The results for the chain B of human p53 are identical as for the chain A of mouse p53 in Fig.4.

(2) In Fig.18 (left: humW & right: humM) the interaction distance R273–DNA (DC15) during 3-ns dynamical changes are shown for the human p53 protein (chain A). For the mutant protein the R273–DNA contact is more close than for the wild-type protein. The results in Fig.18 (human p53' chain A) are correlated with the data in Fig.7 (mouse p35' chain B). In Fig.19 (left: humW & right: humM) the interaction distance R416–DNA (DG13) during 3-ns dynamical changes are shown for the human p53 protein (chain B). For the wild-type protein a minimal distance R273–DNA is around 5 Å, while for the mutant protein a minimal distance R416–DNA increases up to 7 Å. It is obvious that R273H mutation effects a contact distance brtween the p53 human protein with DNA, even though the mutation point is located on chain A which is not close as chain B to DNA. Again the results for the human p53' chain B are correlated with the data for the mouse p35' chain A (compare Figs.19 and 6).

(3) In Fig.20 (left: humW & right: humM) the interaction distance R280–DNA (DC15) during 3-ns dynamical changes are shown for the human p53 protein (chain A). In Fig.21 (left: humW & right: humM) the interaction distance R423–DNA (DG13) during 3-ns dynamical changes are shown for the human p53 protein (chain B). From Figs.20 and 21 we see that the distance distributions between wild-type and mutant proteins are similar. For both chains A and B of the human p53 protein, as like as for the mouse protein (Figs.8 and 9), there no visible effect of R273H mutation induced on distances R280–DNA and R423–DNA can be observed.

(4) In Fig.22 (left: humW & right: humM) the interaction distance S241–DNA (DC15) during 3-ns dynamical changes are shown for the human p53 protein (chain A). The results of Fig.22 are well correlate with data for the chain B of the mouse p53 protein (compare with Fig.11). In Fig.23 (left: humW & right: humM) the interaction distance S384–DNA (DG13) during 3-ns dynamical changes are shown for the human p53 protein (chain B). For the human p53' chain B the S284–DNA distance possesses smallest values among all five amino acid residues of the DNA binding domain. The comparison of Fig.23 with data for the chain A of the mouse p53 protein (Fig.10) are straighforward. As like as in p.(2) point mutation R273H (induced on chain A) causes an effect into oppsotie chain B.

(5) In Fig.24 (left: humW & right: humM) the interaction distance A276–DNA (DC15) during 3-ns dynamical changes are shown for the human p53 protein (chain A). In Fig.25 (left: humW & right: humM) the interaction distance A419–DNA (DG13) during 3-ns dynamical changes are shown for the human p53 protein (chain B). We note that the results for the human and mouse p53 proteins are well correlated (compare Fig.24(chain A)–Fig.13(chain B) & Fig.25(chain B)–Fig.12(chain A)).

11.4. Summary

In summary the effect of single amino mutation Arg273His (R273H) has been investigated for the structures of mouse and human p53 proteins. The study has been aimed on molecular mechanism of the p53 binding with DNA which is important for the physiological functioning of the p53 protein and understanding the origin of cancer disease.

We have examined the structural correlation effects between the p53 mouse (PDB entry: 3EXJ) and human (PDB entry: 1TSR). Four p53 dimer (chains A and B) structures have been simulated and the MD data were compared:
—(1 and 2) The wild-type (mouW) and mutant R273H (mouM) mouse p53 proteins,
—(3 and 4) The wild-type (humW) and mutant R273H (humM) human p53 proteins.

In the DBD (DNA binding domain) three arginines (R248, R273, R280), one serine (S241) and one alanine (A276) are responsible for the p53–DNA binding. The distance distribution for all five key amino acid residues have been estimated and correlation established well between the mouse and human p53 protein structural behaviors.

A comparative analysis show that R273H mutation causes an essential effect on the p53–DNA interaction, removing their close contact. For the mouse p53 (chain A) wild-type protein a minimal distance R273–DNA is around 4 Å, while for the mutant protein a minimal distance H273–DNA increases up to 8 Å. The same observation is true for the human p53 protein.

The MD results show a good agreement with experimental observation, where the R273H mutation has shown simply remove the DNA contact. The obtained statistical data demonstrate detailed conformational changes in the important contact domain DBD between two different, but structurally identical, – mouse and human p53 proteins.

Acknowledgments

The MD simulations have been performed using computing facilities, software and clusters at CICC (JINR, Russia), RICC (RIKEN, Japan), RIKEN-Yokohama (MDGARPE-3), Yasuoka Laboratory (Keio University, Japan). We thank Prof. Toshiaki Iitaka (CAL RIKEN) for the help in getting access to the software and hardware of the RIKEN computing cluster RICC.

Appendix

On four p53-relevant structures as discribed above (the mouse p53 protein: the wild-type and mutant R273H version (PDB entry file "3EXJ") & the human p53 protein: the wild-type and mutant R273H version (PDB entry file "1TSR")), we have performed in total 16 model calculations (periodic PME-NPT; periodic PME-NVT; non-periodic cutoff; and non-periodic no cutoff). All simulations were performed with the AMBER (versions 7 to 11) molecular dynamics software package for studying biomolecules [16-17]. In our periodic MD simulations, the electrostatic interactions were treated with the Particle Mesh Ewald (PME) algorithm [16-18]. Non-periodic MD involved both cutoff and no cutoff simulations; to perform no cutoff calculations the AMBER versions were used which had been adapted under the MDGRAPE-2 and 3 hardwares. For the MDGRAPE-2 and 3, all the particle interactions are calculated [19]. The

Cornell et al. all-atom force field was used in the MD simulations [20-21]. A system was solvated with TIP3P molecules [22] generated in a rectangular and spherical (non-periodic) water baths. The temperature was kept constant by using the Berendsen algorithm [23]. Only bond lengths involving hydrogen atoms were constrained using the SHAKE method [24]. We suppose a comparison of the results of the periodic PME-NPT, periodic PME-NVT, and non-periodic simulations to be the subject of a separate paper. The result of simulations and images of the simulated proteins were analyzed using the RasMol [25], MOL-MOL [26] and VMD (Visual Molecular Dynamics) [27] softwares. For some details of our MD simulations see also [28-30]. Below follows an example of the MD protocol of periodic PME-NPT simulations.

```
************************************************************
Potential function:
    ntf = 1, ntb = 2, igb = 0, nsnb = 1
    ipol = 0, gbsa = 0, iesp = 0
    dielc = 1.00000, cut = 10.00000, intdiel = 1.00000
    scnb = 2.00000, scee = 1.20000
Frozen or restrained atoms:
    ibelly = 0, ntr = 0
Molecular dynamics:
    nstlim = 10000, nscm = 2, nrespa = 1
    t = 0.00000, dt = 0.00100, vlimit = 20.00000
Berendsen (weak-coupling) temperature regulation:
    temp0 = 300.00000, tempi = 0.00000, tautp = 1.00000
Pressure regulation:
    ntp = 1
    pres0 = 1.00000, comp = 44.60000, taup = 1.00000
SHAKE:
    ntc = 2, jfastw = 0
    tol = 0.00001
Ewald parameters:
    verbose = 0, ew_type = 0, nbflag = 1, use_pme = 1
    vdwmeth = 1, eedmeth = 1, netfrc = 1
    Box X = 100.202    Box Y = 68.001    Box Z = 57.148
    Alpha = 90.000    Beta = 90.000    Gamma = 90.000
    NFFT1 = 100    NFFT2 = 72    NFFT3 = 60
    Cutoff = 10.000    Tol = 0.100E-04
    Ewald Coefficient = 0.27511
    Interpolation order = 4
************************************************************
```

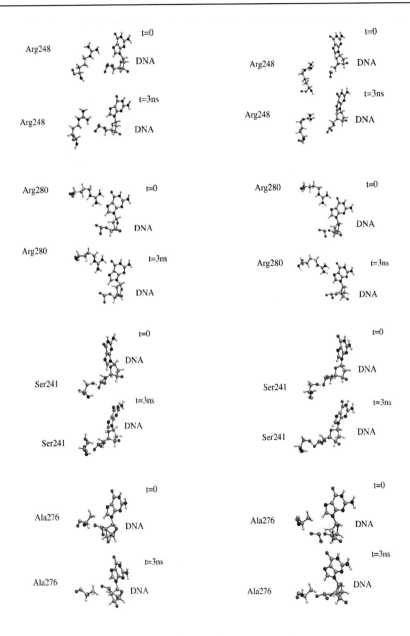

Figure 11.14. The snapshots represent relative amino acid positions {Arg248, Arg273His, Arg280, Ser241, Ala276} → {DNA} for the p53 mouse protein (chain A: wild-type (left); R273H mutant (right)) at t=0 and 3-ns states

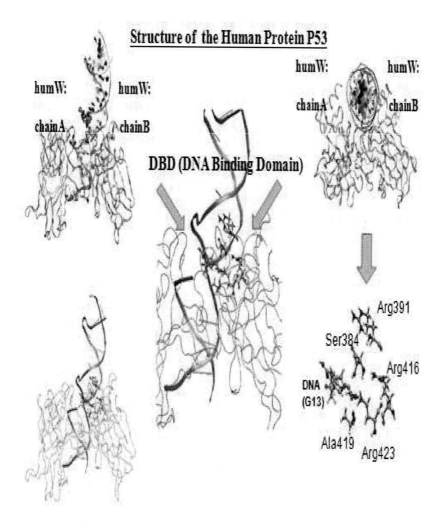

Figure 11.15. The side and top views of the p53 human protein are shown (PDB entry file: 1TSR). For the human p53 protein structure, two chains (A and B) symmetrically surround the related DNA sequence (yellow arrows) located in the central DNA binding domain (DBD).

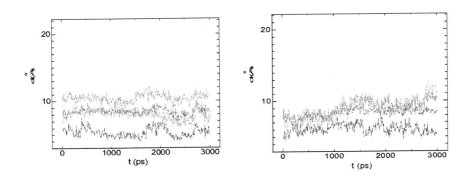

Figure 11.16.   The distance diagrams are presented for the Arg248–DNA (DC15) contact during 3-ns dynamical changes of the human p53 protein (chain A). Left diagram is for the p53 wild-type and right diagram - for the p53 mutant (R273H) protein, respectively.

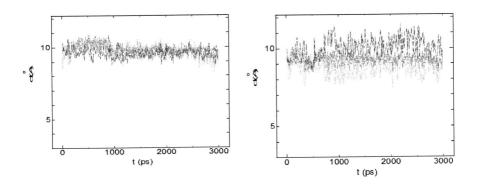

Figure 11.17.   The distance diagrams are presented for the Arg391–DNA (DG13) contact during 3-ns dynamical changes of the human p53 protein (chain B). Left diagram is for the p53 wild-type and right diagram - for the p53 mutant (R273H) protein, respectively.

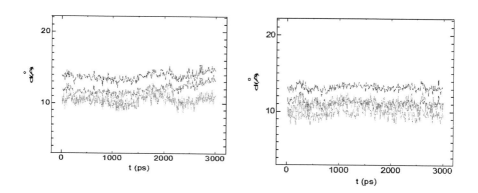

Figure 11.18.   The distance diagrams are presented for the Arg273–DNA (DC15) contact during 3-ns dynamical changes of the human p53 protein (chain A). Left diagram is for the p53 wild-type and right diagram - for the p53 mutant (R273H) protein, respectively.

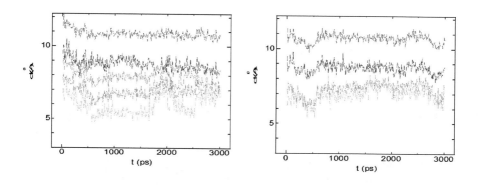

Figure 11.19.   The distance diagrams are presented for the Arg416–DNA (DG13) contact during 3-ns dynamical changes of the human p53 protein (chain B). Left diagram is for the p53 wild-type and right diagram - for the p53 mutant (R273H) protein, respectively.

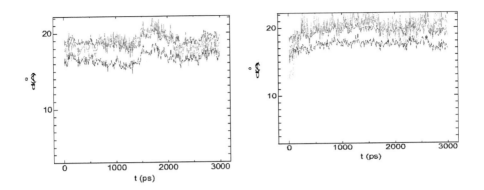

Figure 11.20.   The distance diagrams are presented for the Arg280–DNA (DC15) contact during 3-ns dynamical changes of the human p53 protein (chain A). Left diagram is for the p53 wild-type and right diagram - for the p53 mutant (R273H) protein, respectively.

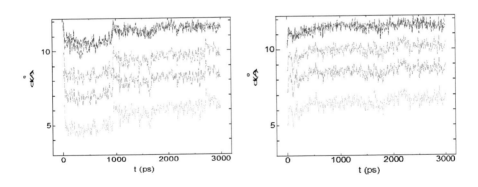

Figure 11.21.   The distance diagrams are presented for the Arg423–DNA (DG13) contact during 3-ns dynamical changes of the human p53 protein (chain B). Left diagram is for the p53 wild-type and right diagram - for the p53 mutant (R273H) protein, respectively.

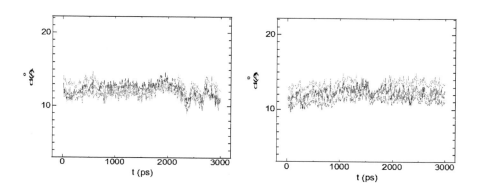

Figure 11.22. The distance diagrams are presented for the Ser241–DNA (DC15) contact during 3-ns dynamical changes of the human p53 protein (chain A). Left diagram is for the p53 wild-type and right diagram - for the p53 mutant (R273H) protein, respectively.

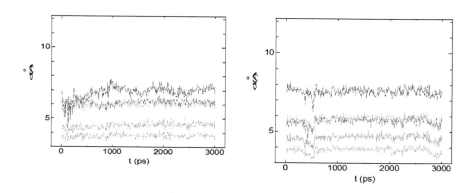

Figure 11.23. The distance diagrams are presented for the Ser384–DNA (DG13) contact during 3-ns dynamical changes of the human p53 protein (chain B). Left diagram is for the p53 wild-type and right diagram - for the p53 mutant (R273H) protein, respectively.

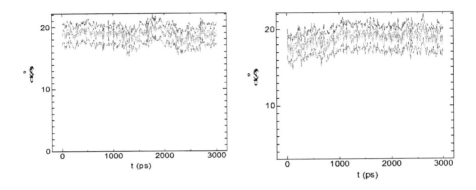

Figure 11.24. The distance diagrams are presented for the Ala276–DNA (DC15) contact during 3-ns dynamical changes of the human p53 protein (chain A). Left diagram is for the p53 wild-type and right diagram - for the p53 mutant (R273H) protein, respectively.

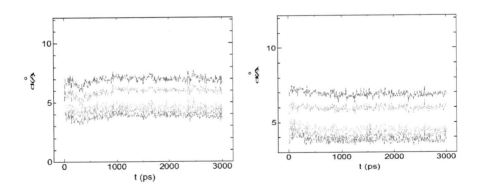

Figure 11.25. The distance diagrams are presented for the Ala419–DNA (DG13) contact during 3-ns dynamical changes of the human p53 protein (chain B). Left diagram is for the p53 wild-type and right diagram - for the p53 mutant (R273H) protein, respectively.

References

[1] Kern S.E., Kinzler K.W., Bruskin A., Jarosz D., Friedman P., Prives C., Vogelstein B., "Identification of p53 as a sequence-specific DNA-binding protein". *Science*, **252**, (1991), (5013): 170811.

[2] Maltzman W., Czyzyk L., "UV irradiation stimulates levels of p53 cellular tumor antigen in nontransformed mouse cells". *Mol. Cell. Biol.*, **4**, (1984), (9): 168994.

[3] Chumakov P.M., Iotsova V.S., Georgiev G.P., "Isolation of a plasmid clone containing the mRNA sequence for mouse nonviral T-antigen" (in Russian). *Dokl. Akad. Nauk SSSR*, **267**, (bf 1982), (5): 12725.

[4] Vinall R.L, Tepper C.G., Shi X.-B., Xue L.A., Gandour-Edwards R. and de Vere White R.W., "The R273H p53 mutation can facilitate the androgen-independent growth of LNCaP by a mechanism that involves H2 relaxin and its cognate receptor LGR7". *Oncogene*, **25**, (2006), 2082-2093.

[5] Joerger A.C., Fersht A.R., "Structural biology of the tumor suppressor p53". *Annu. Rev. Biochem.*, **77**, (2008), 557-82.

[6] Ma B. and Levine A.J., "Probing potential binding modes of the p53 tetramer to DNA based on the symmetries encoded in p53 response elements". *Nucleic Acids Res.*, **35**, (2007); (22): 7733-7747.

[7] Song H., Hollstein M. and Xu Y., "p53 gain-of-function cancer mutants induce genetic instability by inactivating ATM". *Nature Cell Biology*, **9** (bf 2007), 573-580.

[8] Pan Y. and Nussinov R., "Structural Basis for p53 Binding-induced DNA Bending". *The Journal of Biological Chemistry*, **282**, (2007), 691-699.

[9] Lu Q., Tan Yu.-H., and Luo R., "Molecular Dynamics Simulations of p53 DNA-Binding Domain". *J. Phys. Chem. B*, **111**, (2007), 11538-11545.

[10] Zhou Z., Li Y. "Molecular dynamics simulation of S100B protein to explore ligand blockage of the interaction with p53 protein". *J Comput Aided Mol Des.*, **23**, (2009); (10): 705-14.

[11] Wang J., Cao Z. and Li S., "Molecular Dynamics Simulations of Intrinsically Disordered Proteins in Human Diseases". *Current Computer-Aided Drug Design*, **5**, (2009), 280-287.

[12] van Dieck J., Brandt T., Teufel D.P., Veprintsev D.B., Joerger A.C., Fersht A.R., "Molecular basis of S100 proteins interacting with the p53 homologs p63 and p73". *Oncogene*, **8**, (2010), 29(14): 2024-35.

[13] Ang H.C., Joerger A.C., Mayer S., Fersht A.R., "Effects of common cancer mutations on stability and DNA binding of full-length p53 compared with isolated core domains". *J Biol Chem.*, **281**, (2006), (31): 21934-41.

[14] Joerger A.C., Ang H.C., Veprintsev D.B., Blair C.M., Fersht A.R., "Structures of p53 cancer mutants and mechanism of rescue by second-site suppressor mutations". *J Biol Chem.*, **280**, (2005), (16): 16030-7.

[15] http://en.wikipedia.org/wiki/Essential_amino_acid

[16] Pearlman, D.A., Case, D.A., Caldwell, J.W., Ross, W.R., Cheatham, T.E., DeBolt, S., Ferguson, D., Seibel, G., Kollman, P., "AMBER, a computer program for applying molecular mechanics, normal mode analysis, molecular dynamics and free energy calculations to elucidate the structures and energies of molecules". *Comp. Phys. Commun.*, **91**, (1995), 1-41.

[17] Case D.C., Pearlman D.A., Caldwell J.W., Cheatham III T.E., Ross W.S., Simmerling C.L., Darden T.A., Merz K.M., Stanton R.V., Cheng A.L., Vincent J.J., Crowley M., Ferguson D.M., Radmer R.J., Seibel G.L., Singh U.C., Weiner P.K., Kollman P.A., *AMBER*, (**2010**).

[18] Essmann U., Perera L., Berkowitz M.L., Darden T., Lee H. and Pedersen L.G., *J. Chem. Phys.*, **103**, (1995), 8577-8592.

[19] Narumi T., Susukita R., Ebisuzaki T., McNiven G. and Elmergreen B., "Molecular Dynamics Machine: Special-purpose Computer for Molecular

Dynamics Simulations". *Molecular Simulation*, **21**, (1999), 401-408.;
Narumi T., Susukita R., Furusawa H., Yasuoka K., Kawai A., Koishi T.,
Ebisuzaki T., *MDM version of AMBER*, (**2000**).;
Narumi, T., Susukita, R., Furusawa, H., Ebisuzaki, T., "46 Tflops Special-
purpose Computer for Molecular Dynamics Simulations: (WINE-2)".
Proc. 5th Int. Conf. on Signal Processing. Beijing., (**2000**), 575-582.

[20] Ponder, J.W., Case, D.A., "Force fields for protein simulations". *Adv. Prot.
Chem.*, **66**, (2003), 27-85.

[21] Cornell, W.D., Cieplak, P., Bayly, C.I., Gould, I.R., Merz, Jr.K.M., Ferguson, D.M., Spellmeyer, D.C., Fox, T., Caldwell, J.W., Kollman, P.A., "A
second Generation forth field for the simulation of Proteins and Nucleic
Acids". *J. Am. Chem. Soc.*, **117**, (1995), 5179-5197.

[22] Jorgensen, W.L., Chandrasekhar, J., Madura, J.D., "Comparison of simple potential functions for simulating liquid water". *J. Chem. Phys.*, **79**,
(1983), 926-935.

[23] Berendsen, H.J.C., Postma, J.P.M., van Gunsteren, W.F., DiNola, A.,
Haak, J.R., "Molecular dynamics with coupling to an external bath". *J.
Chem. Phys.*, **81**, (1984), 3684-3690.

[24] Ryckaert, J.P., Ciccotti, G., Berendsen, H.J.C., "Numerical integration of
the Cartesian equations of proteins and nucleic acids". *J. Comput. Phys.*,
23, (**1997**), 327-341.

[25] Sayle, R.A., Milner-White, E.J., "RasMol: Biomolecular graphics for all".
Trends in Biochem. Sci., **20**, (1995), 374-376.

[26] Koradi, R., Billeter, M., Wuthrich, K., "MOLMOL: a program for display
and analysis of macromolecular structure". *J. Mol. Graphics*, **4**, (1996),
51-55.

[27] Humphrey, W., Dalke, A. and Schulten, K., "VMD - Visual Molecular
Dynamics". *J. Molec. Graphics*, **14.1**, (1996), 33-38.

[28] Kholmirzo Kholmurodov (Ed.), "Molecular Simulation Studies in Material and Biological Sciences", *Nova Science Publishers Ltd.*, (**2007**),
190p., ISBN 1-59454-607-x.

[29] Kholmurodov, K.T., Hirano, Y., Ebisuzaki, T., "MD Simulations on the Influence of Disease-Related Amino Acid Mutations in the Human Prion Protein". *Chem-Bio Informatics Journal*, **3**, No. 2, (**2003**), 86-95.

[30] K. Kholmurodov, W. Smith, K. Yasuoka, T. Darden and T. Ebisuzaki, *J. Comput. Chem.*, **21**, (2000), 1187.

Index